TECHNOLOGY TRANSFER IN THE DEVELOPING WORLD

TECHNOLOGY TRANSFER IN THE DEVELOPING WORLD

THE CASE OF THE CHILE FOUNDATION

Frank Meissner

Foreword by Julio Luna

PRAEGER

New York
Westport, Connecticut
London

T
174.3
M45
1988

Library of Congress Cataloging-in-Publication Data

Meissner, Frank.
 Technology transfer in the developing world : the case of the
Chile Foundation / Frank Meissner.
  p. cm.
 Bibliography: p.
 Including index.
 ISBN 0-275-92926-4 (alk. paper)
 1. Technology transfer—Developing countries. 2. Fundación Chile.
I. Title.
T174.3.M45 1988
 338.9'26—dc19   87-36127

Library of Congress Catalog Card Number: 87-36127
ISBN: 0-275-92926-4

First published in 1988

Praeger Publishers, One Madison Avenue, New York, NY 10010
A division of Greenwood Press, Inc.

Printed in the United States of America

The paper used in this book complies with the
Permanent Paper Standard issued by the National
Information Standards Organization (Z39.48-1984).

10 9 8 7 6 5 4 3 2 1

To Margit, Paul, and Anne for their patience, and to colleagues in the socioeconomic development profession for stimulating originality via creative imitation.

Contents

Tables

Abbreviations

ACHIGH	Asociación Chilena de Gastronomía
ASACH	Asociación General de Supermercados y Autoservicio de Chile
CAFRA	Cooperativa Agrícola Lechera y de Consumo, Frutillar Alto
CAPRILAC	Comercializadora de Productos Lácteos
CETL	Centro Tecnológico de la Leche
CFI/BC	Crown Forest Industries of British Columbia
CODELCO	Corporación del Cobre
CODESSER	Corporación para el Desarrollo Social del Sector Rural
COMAPAC	Compañía Manufacturera de Papeles y Cartones
CONACYT	Consejo Nacional de Investigación, Ciencia y Tecnología
CONAF	Corporación Nacional Forestal
CONAR	Comisión Nacional de Riego
COPEC	Corporación de Petróleos de Chile
CORFO	Corporación de Fomento
CULTIMAR	Cultivos Marinos Tongoy
ENAP	Empresa Nacional de Petróleo
ENDE	Empresa Nacional de Explosivos
ENDESA	Empresa Nacional de Electricidad S. A.
FAO	Food and Agricultural Organization of the United Nations
FCh	Fundación Chile
FUSADES	Fundación Salvadoreña para el Desarrollo Económico y Social
GoCh	Government of Chile
IBRD	International Bank for Reconstruction and Development (World Bank)
IDB	Inter-American Development Bank
IDLA	Inversiones y Desarrollo Los Andes

INIA	Instituto Nacional de Investigaciones Agrícolas
IMF	International Monetary Fund
INTA	Instituto de Nutrición y Tecnología Alimenticia
INTEC	Instituto Tecnológico/Corporación de Fomento
ITT	International Telephone and Telegraph
JICA	Japanese International Corporation Agency
LDCs	Less Developed Countries
PRO CHILE	Promotora de Exportación de Chile
QCP	Quality Control Program
SERNAP	Secretaría de Recursos Naturales y Pesca
SOTEC	Sociedad Chilena de Tecnología para el Desarrollo
SRI	Stanford Research Institute (SRI International)
USAID	US Agency for International Development

Foreword

Technology is a process by which knowledge and experience are applied to achieving more efficient, effective, and timely use of available resources in a community that aims to increase its cultural and material welfare, according to the community's own values and means. Much of the huge gap in levels of economic development that separates the Third World from industrialized countries is due to open differences in concepts, values, and resources needed to adapt or develop adequate technologies. Some societies consider "industrialized" technologies to conflict with either their moral or intellectual values and aspirations.

Adoption of advanced technology would not automatically close the distance that separates less developed countries from the industrialized world. Yet, within a wide variety of cases and options, the Fundación Chile has characteristics representantive of countries that are at a level of development approaching the industrialized world, or whose economies have already started to reflect important performance indicators. The Fundación Chile model may be useful for Chile. Yet, it might not prove to be fully effective in other countries.

In this book, the author tells the story of how Chile, a country that is approaching the status of a Newly Industrialized Country (NIC) has creatively gone about setting up the Fundación Chile (FCh), an institutions charged with facilitating systematic transfer of relevant technologies, while resolving a bitter dispute with the International Telephone and Telegraph Company (ITT), one of the world's largest TNCs.

The conflict, which eventually gave birth to the Fundación Chile, has been around for a long time. It became dramatized during the government of the Chilean President Salvador Allende. The national telephonic network, was based on a foreign technology derived from industrial development. The network was managed by a foreign monopoly that owned the technology. Chilean politicians, who believed in free enterprise, did not consider nationalization an attractive business for the government to get into. In contrast, defenders of state intervention as catalyst of development, viewed nationalization of the telephonic network as a tool for exercising national sovereignty. The military government, which ended the socialist administration, finally negotiated a compromise solution. To be sure, it does not entirely satisfy any of the contending parties. Yet, the Fundación Chile, born as a "daughter of the crisis", overcame the trauma of its conception, acquired an independent life of its own, and acquired attributes that helped it to contribute to the welfare of the community to which it belongs.

Frank Meissner, a friend and colleague of long standing, has for many years been associated with and committed to development of Latin America. In this book, he brilliantly captures the significance, which the short existence of FCh created for technological development of Chile as well as potential model for sister nations in the Western Hemisphere.

In taking the reader "behind the scenes" of FCh, Dr. Meissner focuses on analysis and interpretation of management and marketing processes that led to conception, birth, growing up and maturing of FCh. He stops frequently along the way to reflect on what the gained experience means in terms of counsel that could be given to assist sister institutions, enterprises and agencies undergoing similar growth pains.

The book balances the "broad-brush" macro-policy issues with the "nitty gritty" micro-project orientation. How are sources of technology identified and projects initiated? What progress has been made towards the goal of financial self-sufficiency? How are FCh projects and pilot enterprises marketed, monitored and evaluated? When do non-performers get weeded out? What lessons does FCh experience hold in store for similar technology transfer and adaptation agencies in Latin America and beyond? Where is FCh likely to be heading for in the second decade of its work?

In the years to come, transnational corporations (TNCs) will undoubtedly remain prime generators and sources of technology for free market economies of newly industrialized countries. Small nations will therefore need to learn how to constructively deal with these giant firms in bringing about creative transfer and prudent adaptation of technology. Development professionals and policymakers as well as knowledgeable laypersons, who want to make the process mutually beneficial, will find much food for thought in Dr. Meissner's book. It therefore gives me a great deal of personal satisfaction to recommend this valuable contribution for all those committed to the development of the Third World.

Julio Luna
Chief, Agricultural and Forestry
Development Division,
Inter-American Development Bank,
Washington, D.C.

Acknowledgments

I became aware of the work of the Fundación Chile (FCh) during the International Symposium at the Annual Convention of the Institute of Food Technologies (IFT), held in Atlanta, Georgia, in June 1981. Two of the symposium papers dealt with FCh and provoked me to find out more about what FCh was doing. My investigations were encouraged by Messrs. Cotton, Flaschen, and Sandvig, three of the founding fathers of FCh. They helped provide background material and opened doors to FCh and International Telephone and Telegraph (ITT). This greatly facilitated the initial gathering of information and its subsequent analysis and interpretation.

The staff at FCh headquarters in Santiago patiently answered my questions and searched their files for the documents I requested. Special thanks go to R. Acklin, Albin Adam, Joaquin Cordua, W.R. Corthorn, Roberto Echeverria, Patricio Galeb, George G. Giddings, Pablo Herrera, Hector Lisboa, Emilio Moreno, Jorge O'Brien, Fernando Sánchez, Juan Pablo Torrealba, and Alfred Vial. I am particularly grateful to M. Wayne Sandvig, Director General of FCh during 1977-85, who answered all my questions. Dr. Anthony Wylie Walbaum, who succeeded Mr. Sandvig in late 1985, made valuable suggestions about the manuscript.

I owe special debt of gratitude to Dr. Luis Alberto Adriasola, who was associated with FCh during its first eight years.

I was given intellectual support by Mr. Julio Luna, my chief at the Inter-American Development Bank. I also want to thank James Dunton and Omar Dahbour, Editors at Praeger Publishers, for their creative suggestions about the form and substance of the manuscript.

I thank my brother-in-law, Dr. Bruno Morawetz, for his moral support and physical sustenance during the writing of this book. My beloved wife, Margit provided many helpful hints and comments.

The biggest debt of gratitude is owed to Ms. Silvia Echeverría, executive secretary at the Inter-American Development Bank, who patiently typed the manuscript and made numerous corrections. Mrs. Barbara Rietveld, editor in the Inter-American Development Bank, prepared the index.

I am, of course, solely responsible for errors of fact, for interpretations of events as well as all conclusions and recommendations for roads ahead.

TECHNOLOGY TRANSFER IN THE DEVELOPING WORLD

1
Introduction

To creatively imitate is the highest form of originality.
Jean-Baptiste Voltaire

THE MEANING OF TECHNOLOGY TRANSFER

Transfer of technology requires a large dose of creative imitation.[1] It takes time and money to generate new technology. Developing countries cannot afford to do basic research and development (R&D). If they want to procure needed technology, they must buy, borrow, steal, imitate, copy, adapt, or beg. The process is called transfer of technology. Few developing countries are good at it. They frequently end up with expensive and irrelevant or inappropriate technologies, which tend to hinder rather than help their socioeconomic development efforts.

Since the mid-1960s much has been written on the nature and impact of the transfer of technology to less-developed countries (LDCs). The socioeconomic development profession initially believed that the opportunities offered by availability of more advanced technologies would vastly simplify and accelerate the process of growth. It seemed logical that gunpowder or the wheel would only have to be invented once. Thereafter anybody could acquire the technology by imitation.

It is now clear that these sanguine expectations were highly naive. Technology transfer is a complicated and costly process. The ultimate success depends on the level and direction of indigenous technical efforts as well as the institutional setting in the recipient country.

In theory technology is conceived as information necessary to design and produce a given good by any number of alternative methods. This information is assumed to be in the public domain freely available, costlessly reproducible, codified in designs and operating manuals.

In reality technology is a quantum of knowledge retained by individual teams of specialized personnel. This knowledge in design, production, and investment activities, is mostly tacit. There are no collections of clear cut blue-prints and manuals. Technology is acquired in problem-solving and trouble-shooting activities within the firm. The know-how is often not proprietory. Yet, other enterprises do not have easy access to it. On the contrary, in each individual firm progressive accumulation of the technical knowledge takes place through a time-consuming and expensive process of learning-by-doing.

For the purposes of this book, technology is defined as the configuration of processes, plans, techniques, knowledge and skills required to effectively produce, process, and market a product or service. Transfer of technology is the act of sharing know-how by such devices as consultancy, joint ventures, gifts, licenses, franchises, and patents.

Technology comes in various forms and is transferred in many different ways. Its benefit may bear no relationship to its newness. A tractor may revolutionize an agricultural process; a second tractor may double the effect. An improved variety of rice may revolutionize agriculture, permanently change cropping patterns, redefine political and social relationships, alter the landscape and even the climate and, in short, create a "Green Revolution."

A great deal of technology is transferred and adapted by demonstrations, dissemination, training, consultancy and management assistance. Individual transactions may be complex, yet these processes all focus on the identification of the required technology, followed by buying and selling.

Knowledge knows no boundaries. The inventor or originator controls technology only so long as it is his alone. Its economic value, when embodied in a productive system, determines its prices. Once technology is shared, control is lost. The proprietor may therefore ask as high a price as is determined by its ultimate potential for increasing productive capacity.

Technology is either homemade or acquired. Few enterprises can hope to be self-sufficient in technology. To do so would mean turning one's back on the contributions of others and wastefully duplicating their efforts. When technology is shared, progress occurs more quickly and surerly. Some social and economic changes may have to take place before technology can be effectively transferred. Sources of indigenous raw materials may need to be developed, human resources trained, policies established, and incentives provided.

The literature of the 1960s and 1970s is full of conceptual discussions of macroeconomic problems related to the generation of production, processing, and marketing technologies and their dissemination and transfer.

In the late 1970s the focus started to shift away from macro to micro issues. An increasing number of writers now tend to analyze decision making about the transfer and absorption of technologies at the level of individual enterprises. They focus on the inevitable and often messy web of contradictions, conflicts, inconsistencies, missing information, misunderstood data, and illogical explanations. The World Bank, the Inter-American Development Bank (IDB), the Foundation for Multi-national Management Education (FMME), and other development agencies are pouring out case studies of technology transfer within countries, industries, sectors, and regions.

The Importance of the FCh

Few of the writings cover the management and marketing aspects of institutions and agencies directly and explicitly involved with technology transfer and adaptation. That is where the Fundación Chile (FCh) comes in.

FCh was created by Decree 1528 issued on August 3, 1976. The purpose of FCh is to facilitate processes by which Chile, a developing country, would gain access to technical knowledge that has been accumulated over generations by developed nations. FCh operates on the premise that, amazingly enough, much of this knowhow tends to be available for the asking, not just from one but from several and often competing sources.

While the number of published case studies is growing by leaps and bounds, the pragmatic emphasis of FCh on learning by doing and readiness to share experience is so far unique. Few studies have exposed these agencies to an open and frank examination of lights and shadows, including autopsies of projects that never left the launching pad and were abandoned. This book shows how FCh adapted customary International Telephone and Telegraph (ITT) operational planning to the politico socioeconomic circumstances encountered in Chile, and how feedback from experience is instantly utilized for improvement of ongoing work.

In short, this book takes the reader "behind the scenes" of FCh. It focuses on analysis and interpretation of management and marketing processes that led to conception, birth, and maturing of FCh. I stop frequently along the way to reflect on what the experience gained means in terms of counsel that could be given to assist sister institutions, enterprises, and agencies who are undergoing similar growth pains.

Purpose of This Book and Sources of Data

The prime purpose of this book is to show interested readers—entrepreneurs, economic and technology policymakers, development and management consultants, inventors, researchers, academicians, students of development and informed lay persons as well as FCh insiders—what FCh has been doing, why, and where it is heading. The story is told "as is." This is because I feel that FCh can serve as model for those willing to learn from experience rather than insisting on repeating costly errors.

In arranging the book I attempted to balance the "broad brush" macro-policy issues with the "nitty-gritty" micro project orientation, which would make it difficult for the reader to see the forest for the trees. Consequently in the core of the text I focus on the essence behind this curious marriage of the government of Chile (GoCh) with ITT. Who brought the partners to the altar? Who governs FCh? What policies guide identification and selection of opportunities for technology transfer and adaptation? How are sources of technology identified and projects initiated? What progress has been made towards the goal of financial self-sufficiency? How are FCh "cases" (projects) monitored and evaluated? When do non performers get weeded out? What lessons does the FCh experience hold in store for similar technology transfer-and-adaptation agencies in Latin America and elsewhere? Where is FCh likely to be heading in the second decade of its work?

Profiles of individual projects appear in the core of the text.

For inexperienced outsiders the case reports might appear pedestrian. So be it. The experienced insiders might see in them repeated touches of hardworking imaginative problem solvers. To them this book is dedicated.

I am an "old Latin America hand" who has observed FCh from its inception. Since the mid-1970s I got to know many actors in the FCh joint ventures: FCh professional staffers, ITT executives and their Chilean counterparts, FCh clients and consultants, as well as outside kibbitzers.

In writing the descriptive part of the book, I have relied on primary sources of internal data, including Plans of Operation for 1982, 1983, 1984, and 1985; performance monitoring reports and cost/benefit projections provided to the Board of Directors; preliminary, pre-feasibility and feasibility studies for individual projects; press releases prepared by FCh; and reprints of media reports dealing with FCh activities.

Most of the financial data in these documents appear in current or adjusted Chilean pesos. With a rate of inflation in the low 2 digits these figures mean little to most readers. I therefore tried to convert the data into U.S. dollar equivalents. This type of dollarization undoubtedly raises some eyebrows for methodology-conscious readers. To them I apologize beforehand, and reassure them that data are presented strictly for illustrative purposes, merely to show orders of magnitude. The dollar conversion of 1985 budget figures and costs/benefit evaluations in Chapters 4 and 5 have been particularly bothersome. To explain why rows and columns do not balance would have called for reams of footnotes, an effort not warranted within the framework of this book. I have received assurance from FCh management that any bona fide individual, interested in methodology, is welcome to approach the Foundation for explanations.

In researching for this book I have been given free access to management and staff of FCh, obtaining straightforward answers to all questions asked. I never felt that information was intentionally being withheld. Often I undoubtedly failed to ask the right questions.

FCh is continuously changing. Today's facts are obsolete tomorrow. Officially born on August 27, 1976, FCh was managed by U.S. chief executives until December 31, 1985. In January 1986 management passed into Chilean hands.

In a letter of August 14, 1987, Dr. Anthony Wylie, Director General of FCh, pointed out to me that in the case of jojoba, following a considerable debate,

> we [FCh] reduced our effort as there were numerous other institutions which in theory at least were conducting similar work. Now, two years later, little has happened in this area... We would not like this to be interpreted as a failure on our part, as it is something we stopped pushing.
> Another case is the goat cheese project, in which we got the product established on the market and reached one of our initial objectives, convincing people this was a good and safe product. Subsequently, however, for lack of supply of goat milk, we decided to close the plant in Ovalle.

Dr. Wylie suggested that I might want to include the following caveat somewhere in the text: "The information contained in this document refers to some of the projects on which Fundación Chile worked during its first 10 years of operation (1976-85). As could be expected some of the [projects] have subsequently been either substantially modified or discontinued as the conditions in which they were conducted changed."

I have followed this useful advice and entirely or partially left out analyses of these "dropout." However, since I have written up those cases anyway, copies will be provided, at cost, to curious readers who request them.

Scope and Method of Presentation

The core of this book consists of nine chapters. Chapter 1 introduces the topic, defines technology and explains how the book came about. Chapter 2 is a review of the conception and birth of FCh, its objectives, strategies, and evolution into its current organizational and financial structure, including staffing.

Chapter 3 is the core of the book. It explains how FCh plans and monitors its activities within the framework of 10 (ten) objectives, which are explicitly spelled out in the 1985 Plan of Operations: (a) create subsidiary enterprises demonstrating feasibility of using specific technologies under conditions prevailing in Chile; (b) provide technical assistance aimed at increasing the productivity of specific types of industries, fisheries, and agriculture; (c) improve domestic marketing of selected products; (d) develop nontraditional exports; (e) promote and implement quality control and certification for selected Chilean export products; (f) encourage production of nontraditional high unit-value crops for export; (g) market technology transfer activities of FCh; (h) improve management of FCh subsidiaries; (i) expand FCh outreach beyond borders of Chile; and (j) generate funds needed for funding science- and technology-oriented enterprises.

Chapter 4 provides insights on how FCh approaches the task of gradually achieving self financing. Chapter 5 reviews methodology used in monitoring FCh performance and measuring cost/benefit impacts of its activities. It contains a great deal of conceptual food for thought.

Chapter 6 focuses on ways in which FCh markets itself, an idea subject to benign neglect by most FCh sister institutions. Development of marketing functions within FCh are traced, methods of FCh reach out are enumerated, including creation of the Chilean Society for Technology Development (SOTEC), launching of the Agricultural Information System, and preparation of publications, technical short courses, seminars, and workshops.

Chapter 7 summarizes the experience of the first decade of FCh activities. Chapter 8 draws conclusions about dos and don'ts derived from an outsider's analysis of the numerous activities FCh has engaged in. Chapter 9 focuses on the agenda for the second decade of FCh. Chapter 10 provides an epilogue, which updates the political and macroeconomic climate of privatization and free market economy within which Chile has operated since assumption of the presidency by General Augusto Pinochet in 1973.

The Bibliography presents a list of selected references, focusing on publications sponsored by FCh or about FCh, as well as writings directly related to transfer of technology. Considering the ongoing flood of literature on the subject, a listing of that sort becomes instantaneously obsolete. The reader is advised to do his own update.

The setting for each major topic is established by quoting statements made by persons or appearing in documents related to the subject at hand as well as in songs, proverbs, maxims of conventional and nonconventional wisdom. This is to add flavor to the text as well as to facilitate its reading.

Most of the topics in this book have already been discussed in some publication or another. Most of the raw data are therefore in the public domain. However, the interpretation of the data is strictly mine. I am obviously an outside observer, who has not been privy to the "in-house" deliberations, which swung decisions about individual projects in one direction or another. I therefore try merely to explain FCh to other outsiders who are interested in what FCh has been doing and the alternative roads it may take in years to come.

Contemplated Follow up

Ten years is a short time in the life of an institution with such mid- and long-term objectives as transfer and adaptation of technologies. It is therefore too early to make broad philosophical and conceptual pronouncements about the track record of FCh. The individual reader will need to decide to what extent the experience is relevant, and how specifically to pick FCh brains for her/his own benefit.

Furthermore, learning is an ongoing process. Therefore let me finish this introductory section with a plea for feedback. What flaws in the methodology, used in analysis of the track record of FCh's work, can be detected? What factual errors? What improvements should be made so as to strengthen institutional memories within FCh? What measures should be taken so as make this experience a more useful prelude to future work? How should FCh reach out to sister institutions throughout the world, so as intensify cross-pollination of ideas and exchange of experiences?

I would be pleased if the book were to inspire other researchers to write up similar case studies for FCh sister institutions. Such documents would help establish the backlog of experience from which the development profession might eventually be able to derive some basic rules about successful management and marketing of agencies concerned with technology transfer.

NOTES

1. The essay deals with Chile, a "developing country", and ITT, a "transnational" corporation. Let me clarify some definitions of those words.

The expression transnational corporation (TNC) is used in preference to such alternatives as multinational or international. This is because ITT's corporate activities are coordinated across nations in accord with global strategies rather than merely operating in many host countries.

"First"," Second" and "Third World Nations" (TWNs) as well as Less Developed Countries (LDCs) will be used in referring to a broad set of nations with different levels of per capita income and industrial development. The term "Third World Nations" stresses the attempts to find development paths necessarily different from the First World (North America, Western Europe, and Japan); and the Second World (Soviet Union and Eastern Europe). The term "developing nations" suggests that real output per capita is rising in these countries. The term 'Less

Developed Countries (LDCs) indicates lower levels of per capita product than in more-developed countries.
2. Nathan Rosenberg and Claudio Frischtak, <u>International Technology Transfer</u>, New York: Praeger, 1985, p. vii.

2
Origins, Structure, Staff, and Operations of FCh

The Fundación Chile is a joint venture between a government and a transnational corporation aimed at systematically incorporating new technologies into a country's economy. That makes it unique not only in Chile but in the entire world.

M. Wayne Sandvig, Director General of Fundación Chile, 1977–85

ORIGINS

In explaining how the country was created, Chileans like to tell a variation on the Biblical story of Genesis. It goes something like this. When God created Earth, a few bits were left over: (a) morsels of desert; (b) a few luxuriant forests; (c) enormous mountains; (d) odd pieces of fertile valleys, lakes, ice, snow; (e) a bevy of islands and beaches; (f) lots of sea, and (g) a small central plain. God then looked for an isolated spot, and deposited the whole lot there, carefully and tenderly, like an artist putting finishing touches on a picture.

That is allegedly how Chile became a mosaic of landscapes: burning deserts of the north; furious waves of the Pacific Ocean to the west; rainy forests and ice fields of the Antartic in the south; and the Andean mountains to the east.

Chile is 4,200 kilometers long with an average width of 180 kilometers. There is no other country in the world so tightly squeezed between the sea and the mountains. That is why on maps Chile looks like a sword in the back of the South American continent; an Andean balcony looking out over the Pacific.

Chile is a beautiful country. Richly endowed with natural and human resources it has long been attracting investors.

Included among the latter was the International Telephone and Telegraph Company (ITT), which set up a subsidiary called the Chilean Telephone Company (CHILTECO).

During the 1970-73 regime of President Salvador Allende Gossens numerous industries and banks were nationalized. Former owners were to be compensated over a period of 30 years, with 3 percent interest paid on the book value of the property. When the government decided to nationalize CHILTECO; it offered a compensation of $24 million for the 70 percent of shares ITT held in CHILTECO.

ITT claimed that the book value of its expropriated properties in Chile was $153 million. The risk insurance, held with the U.S. Overseas Private Investment Corporation (OPIC), would have covered only $90 million. ITT stood to lose at least $50 million. Negotiations about ways to reduce the loss went on and on and on.

In 1974, when the impasse appeared to have no solution, Raúl Saez, then Minister of Economic Coordination, was visiting New York. A simple idea struck him in a moment of creativity. Let ITT and the Government of Chile (GoCh) split the difference on the disputed $50 million, and devote the resources to a mutually beneficial activity. What activity? Chile, a country bountifully blessed with natural resource for agriculture, fisheries, forestry, and mining needed ways to use these resources more effectively. ITT already had a successful technology laboratory in Spain. Could something similar be done in Chile? Perhaps not as a subsidiary but a true long term partnership with GoCh putting up $25 million as matching counterpart to the $25 million of ITT.

Raúl Saez tried that sketch of an idea on Eugene Black, the recently retired president of the International Bank for Reconstruction and Development (IBRD), generally known as the World Bank. Mr. Black liked the concept and promptly made an appointment for Minister Saez to meet with Harold Geneen, president and chief executive officer of ITT. Mr. Geneen agreed that this would be constructive approach to getting out of a bad situation. Steward S. Flaschen, an executive in ITT's Latin American operations, was promptly sent to Santiago. His mandate was to raise the joint venture flag to see whether Chilean authorities would salute. They did.

Following lengthy negotiations the concept skeletons were fleshed out. By Legal Decree 1528, issued on August 3, 1976, GoCh recognized the Fundación Chile (FCh) as a private, nonprofit organization and approved its by-laws and operating statutes. The joint venture of GoCh and ITT was in orbit. It was aimed at "promoting the transfer of new technologies, methods or systems which can contribute to development of productive activities in the country."

The initial endowment of $50 million was made available to FCh according to the following schedule: initial 3 years, 1976-78, $8.0 million annually; $4.0 million annually for the subsequent 6 years, 1979-84; and the remaining $2 million in 1985.

These funds were being supplemented by service fees, which gradually were to become ever more significant sources of income. In 1984 FCh revenues from sale of goods and services were the equivalent of 30 percent of operating costs. The Board of Directors stipulated 50 percent self support to be one of FCh's goals for 1986/87.

GOVERNANCE

Mutual prejudice tends to characterize relations between private transnational corporations (TNCs) and public sector institutions in developing countries. TNCs tend to harp on perceived as well as real inefficiencies and mismanagement of government agencies and enterprises. By the same token governments in less developed countries (LDCs) have their own implicit or explicit biases against private enterprise: They tend to view TNCs as lacking in social values, devoid of national spirit, intruders on the sovereignty of host nations. Many LDCs therefore believe that TNCs need to be harnessed and controlled so as to protect the welfare of their societies.

FCh is a successful marriage of two unlikely partners, a rare case of "making a silk purse out of a sow's ear." This is because FCh was conceived under circumstances so controversial and delicate that most writers prefer to skip the subject altogether.

Establishment of FCh did not mean that perceptions that the partners had of each other changed suddenly and completely. The founding fathers of FCh were fully aware of the need to create an active Board of Directors that would perform more than the normal functions of establishing policies of FCh and supervise its labors. They keenly felt that the Board, rather than being merely a rubber stamp for GoCh and ITT decisions, had to be an active forum conducive to generating constructive reconciliation on potentially divisible issues. At the same time the Board needed to encourage creative ideas about what specific purposes FCh should serve and how to go about implementing them. So as to build that bridge of mutual understanding the right to appoint members to the Board was divided equally between the partners: 5 members (called consejeros, or advisors) and 5 alternates are named by GoCh and by ITT. One of GoCh members was to become president, and one of the ITT members vice-president, of the Board.

Since FCh's inception in 1976 the president of the Board was General Manuel Pinochet Sepúlveda (no relative of General Augusto Pinochet, President of Chile), who heads the National Corporation for Manufacture of Explosives (Empresa Nacional de Explosivos, ENDEX).

ORGANIZATION

Initial organization of FCh was preceded by an inventory of natural and human resources plus a thorough survey of relevant government, university and private industry groups. Many respondents of that initial survey were co-opted to make up a broad support network for FCh, consisting of "natural" leaders in business, industry, finance, labor, agriculture, and academia, as well as government.

The first organization chart was simple. A Director General responsible for overall program development and operations; two deputy directors: one for marketing and development of Food Technology, the other for Finance and Administration. As FCh grew the organization chart became more complex. As of December 31, 1985 FCh was composed of 6 main departments: General Management; Food Technology; Forestry; Marketing; Finance and Administration; and Electronics and Telecommunications. In addition, FCh had an informal international Technology Search System (TSS) to satisfy its own needs and those of third parties.

The 5000 m2 FCh headquarters complex in Santiago houses chemical, microbiological, and microelectronic laboratories, a pilot plant for food processing, a technical library and staff offices. In addition FCh ran 5 associated enterprises: 2 related to fisheries, one cheese processing plant, a beef packing operation, and a vertically integrated berry production packing export marketing enterprise.

HUMAN RESOURCES

The initial staff of FCh consisted of 5 experts from overseas (the Founding Party): Dr. Robert Cotton, a senior ITT food industry research and development executive, was named the first director general of FCh. He was joined by a food technologist, a nutritionist, a chemical engineer with background in U.S. Department of Agriculture (USDA) and food industry, and an ITT telecommunications specialist. A Chilean national, with financial experience at ITT and Exxon units in Chile and abroad, was named deputy director.

Early in the game a crucial decision mandated that Chilean nationals were gradually to take over FCh activities. A vigorous recruitment of qualified Chileans was carried out so as to fill upper echelons of management. In house training programs were provided to prepare capable candidates for key positions of leadership and responsibility.

As of January 1, 1986 FCh had a permanent staff of 86 at headquarters in Santiago, plus approximately 150 more workers in the subsidiary enterprises. National and international consultants supplement the staff in areas of specialized knowledge.

The 60 professional staffers had undergraduate degrees in the following fields of specialization:

Table 2.1 Professional Staff of FCh, 1986

	Number
1. Agriculture and Food Technology	22
2. Engineering:	16
2.1 Civil 4	
2.2 Electrical 5	
2.3 Industrial 2	
2.4 Chemical 5	
3. Accounting	6
4. Business and Public Administration	6
5. Veterinary Medicine	3
6. Forestry	2
7. Fisheries	2
8. Interpreters	3
Total	60

There were 17 postgraduate degrees holders, including three Ph.D.s.

Wayne Sandvig, who was second Director General 1977–85, came to Chile from Madrid (Spain), where he managed ITT's new product development laboratory for electronics and telecommunications.

Continued basic development and training of these highly qualified professionals, a permanent concern of FCh management, is being undertaken in at least three ways.

(a) Refresher courses and study in centers of recognized prestige in Chile and abroad. In 1984 alone over 20 FCh professionals spent 200 working days on trips outside the country.

(b) In exploring possibilities for adaptation and/or development of related sets of technologies, FCh tends to reinforce its own team of experts with consultants. They are drawn from national and foreign technology centers, as well as from Chilean universities. In 1984 alone FCh used an equivalent of 400 expert/working days provided by 33 foreign specialists. The consultants became FCh alumni, making up part of the world wide outreach network.

(c) FCh also organizes inter institutional groups to undertake specific tasks. The preferred mechanism are work teams made up of FCh's professionals and national or foreign consultants.

OPERATIONS

During the first decade of FCh's operations the Chilean economy tended to be plagued by a series of constraints: huge foreign indebtedness; high rates of interest in international capital markets; low prices for copper, the traditionally the main Chilean export commodity; large distance from major overseas markets; domestic decapitalization, and high taxation and saving rates; imperfect domestic capital market; psychological reluctance of Chilean entrepreneurs to assume risks related to mid-and longer term investments; inadequate coordination of export activities; and lacking incentives for investment in productivity improvements, including transfer of technology.

In the 1985 Plan of Operations, FCh executives explicitly pointed out that countries with fewer natural resources than Chile have achieved spectacular economic progress. Due to its excellent university as well as technical educational system, Chile developed plentiful human resources that tended to remain underutilized. FCh was convinced that that brain drain of the best professional talent could be reversed into a brain gain by implementing a systematic exploitation of available natural resources in agriculture, fisheries, forestry, mining, and electronics.

The key government strategy for achieving these objectives consisted of export promotion of goods and services, in which the country could combine comparative advantage with adequate technology. From its inception FCh was therefore to be merely one of the catalysts that would help match resources with required know-how, aiming at effective exploitation of Chile's resources.

NOTE

1. The history of ITT in Chile is reviewed in Chapter 11 of Anthony Simpson, ITT: The Sovereign State, London: Holder and Stoughton, 1973.

3

Profiles of FCh Projects

Fundación Chile was officially created on
August 3, 1976 by decree-law 1528. It's
mission is to serve as a 'bridge' for transfer to the country of new technologies, which
will contribute to a better use of its
natural resources.

Fundación Chile, 1985 Annual Report

The partial track record of FCh's achievement, presented in
this chapter, starts with the establishment in 1976 of the
initial $50 million endowment, $25 million from the Chilean
government, and $25 million from ITT. The decade ends with the
appointment, on January 1, 1986, of Dr. Anthony Wylie Walbaum, a
Chilean citizen, as Director General of FCh, taking over from M.
Wayne Sandvig, an ITT executive and U.S. citizen who directed FCh
during 1977-85.

The FCh Annual Report for 1985 indicates that in the First
10 Years (Los Primeros 10 Años, Annex 2) at least 86 major events
took place in a wide variety of activities: agroindustry, marine
resources, forestry, formulation of new food products,
electronics, telecommunications, quality control, nutrition and
processing of cereals.

In the first 5 year period 28 events occurred, more than
doubling to 58 in the 1981-85 period:

Table 3.1 Major FCh Events, 1976-85

Year	Number	Year	Number
1976	3	1981	10
77	4	82	11
78	6	83	13
79	7	84	13
80	8	85	11
Subtotal	28	Subtotal	58

Source: Appendix 1

The events listed above are merely to show the diversity of areas and activities to which FCh has brought new technologies. Several types of activities are repeatedly referred to; the technical cooperation assistance to the canning industry (CAT) is mentioned 4 times, even though it has since been abandoned; the successful Quality Certification Programs (QCP) appear 5 times; the PROCARNE packing house for vacuum-packed, boxed beef, 4 times. Yet not mentioned are such important events as establishment of the management unit for subsidiary enterprises, irradiation of smoked salmon, 1985 reorganization of FCh, and so on.

This type of random enumeration of activities evidently does not serve as solid basis for reviewing the management style of FCh. After all, the main task of any institution, concerned with technology adaptation and transfer, is to identify ideas, separate grain from chaff, and gradually sift the remaining projects through nets of preliminary study, pre-feasibility and feasibility studies, final design, implementation, and ex-post evaluation.

It would have been completely unwieldy to go through the development trajectory of all the projects FCh has undertaken during its first decade of activities. Which projects should be selected and how much should be said about them?

The first question is readily answered: The 1985 FCh Plan of Operation listed about 30 ongoing projects, or "cases" as they are referred to in the ITT jargon. That seemed to be a practical point of departure. Furthermore, the projects were pre-arranged into four major groups.

The first group contained 10 projects that had already reached or were close to reaching the status of subsidiary enterprises. These enterprises aimed to provide real-life demonstration of the overall socioeconomic viability of the technologies FCh has helping to adapt and/or transfer. In some of those enterprises FCh already had partners, others were on the way to becoming joint ventures. None of them had yet reached the ultimate stage: FCh selling its stake in the enterprises to

Chilean and/or foreign investors, and using proceeds for setting up new technology based ventures.

The second group focused on projects of FCh technical assistance to clients interested in improving technologies related to production, processing, and marketing specific goods and services.

The third group contained 4 variations on the theme of infrastructure for promotion of nontraditional exports: FCh's growing Quality Certification Programs (QCPs) in fruits, fisheries, seafoods, wine, and wood.

The fourth group consisted of two areas related to ITT's prime line of technologies: microprocessors and rural telephony.

Marketing of technology, including transfer of marketing technologies, is a crucial and unique components of FCh work. It is reviewed separately in Chapter 4.

Table 3.2 will provide the reader with an introductory orientation to the 4 major groups, under which the individual projects appear, including: (a) identification of the primary source of the technology, (b) degree of adaptation done to make the technology applicable to Chilean conditions, (c) primary target markets, (d) potential profitability, and (e) status of the project as of January 1, 1986.

The question of how much detail to go into concerning each project is a great deal more difficult. If any valid macro conclusions are to be drawn about application of FCh's management style, they must be based on empirical reality, not armchair speculation. There are no aphorisms that can adequately explain the context within which decisions about the fate of individual projects were made.

If this book is to be useful for practitioners and policymakers then individual project profiles must take the reader step-by-step through different stages involved in adaptation and transfer of technology: (a) definition of problem to be resolved or challenge to be faced; (b) search for sources of suitable technologies; (c) making technologies location specific; (d) marketing resulting technology hardware or software to potential implementors; (e) evaluating the socioeconomic impact of those technologies; and (f) if relevant, exploring alternative follow-up strategies.

The project profiles differ in length and detail as well as in form of presentation. This reflects the quality of the reference materials consulted at FCh. The good and the bad news, related to current status and outlook for individual projects, are faithfully presented and interpreted. To tell this story adequately calls for review of some unglamorous nitty-gritty details. This will undoubtedly appear pedestrian to many broad-brush conceptualizers in the academic world, and perhaps even to some LDC macro-economists and policymakers responsible for creating investor friendly environments. But in the last analysis, individual investors and entrepreneurs are micro-economists. This is because eventually, a specific

Table 3.2 Four Major FCh Types of Projects

| Project | Technology | | Market Target b/ | Status as of January 1986 c/ |
	Primary Source	Degree of Required a/ Adaptation		
1. Technology Oriented Enterprises (TOE)				
1.1 Salmon Farming and Ranching	US/Canada	SC	E	FE
1.2 Oyster Cultivation	Japan	SC	E	FE
1.3 CAPRILAC Goat Cheese	France	SI	L	A
1.4 PRO-CARNE Boxed Beef	US	SI	L	QO
1.5 Berries La Unión	Swiss	SI	E	FE
1.6 Forestry Related Enterprises	US/Canada	SI	L/E	SP
1.7 Utilization of Crop Waste	US/Canada	M	L	QO
(a) Forestry				
(b) Fruit				
(c) Livestock				
1.8 Management of TOEs				
2. Technical Assistance Services				
2.1 Asparagus	US	SI	E	FE
2.2 Dairy Related Technology				
(a) Products	Swiss	M	L	FE
(b) Small Cheese Plants	Swiss	M	L	FE
(c) Dairy Livestock Management	US	M	L	FE
2.3 Control of Ulex Infestation	New Zealand	SC	L	PE
2.4 Tenderising Dates	Israel	M	E	E
2.5 Fisheries Related Technologies	US	M	L	E
2.6 Food Irradiation	US/South Africa	M	E	QO
2.7 Jojoba Bean	US	SC	L	QO
2.8 Minced Fish	US	M	L	FE
2.9 Lupine	Germany	M	L	QO
2.10 Canning Industry	US	M	L	A
2.11 Agro-Economic Information System	US	SI	L	E
2.12 Kiwifruit	New Zealand	M	E	FE
3. Quality Certification Program (QCP)				
3.1 Fruits	US	SI	E	FE
3.2 Sea Foods	US/South Africa	M	E	FE
3.3 Wine	US	SI	E	E
3.4 Wood	US/Canada	SI	L/E	E
4. Electronics and Telecommunications				
4.1 Microprocessors	US(ITT)	M	L	FE
4.2 Rural Telephony	US(ITT)	M	L	FE

a SC = substantial changes in selected technology undertaken, requiring extensive location, specific research and field testing of resulting adaptations.

SI = selected technologies readily adaptable, while production of raw material, processing and marketing require systematic integration

MA = selected technology usable with relatively minor adaptations

b E = Export, L = Local

c A = Abandoned, FE = Firmly Established, E = Established, PE = Preliminary Evaluation, SP = Successful Pilot operation carried out, QO = Questionable Outlook for success within next 5 to 10 years.

Estimates of socioeconomic benefits, generated by the above projects, appear in Chapter 5, together with explanation of methodology used in making these appraisals. Also included are cash flow projections for 1984-90 based on September 1984, whenever such data were available.

risk-taker in the public and/or private sector must make a decision at the level of a specific firm, about a specific set of actions related to transfer and adaptation of a specific type of technology from a specific source.

Several alternative ways to present Chapter 3 were considered: (a) include the project profiles in their entirety; (b) relegate most of the project profiles to an annex, and dedicate the body of the text to analysis and conclusions; or (c) leave out some of the project profiles entirely, and make a collection of them available to readers on request to the author. For the sake of brevity the third alternative was adopted. The removed profiles deal with enterprises or activities that—due to unexpected changes of circumstances—have been substantially modified or discontinued altogether: 1.3 Caprilac, 2.6 Food irradiation, 2.7 Jojoba beans, 2.9 Lupine, and 2.10 Canning industry.

TECHNOLOGY-ORIENTED ENTERPRISES

One of the key commandments of FCh is to prove the viability of adapted technologies. This is being done by creating advanced technology-oriented private commercial enterprises. Consequently, from its very inception in 1976 FCh focused much of its attention and energy on demonstrating that it believes in what it is preaching. FCh has therefore been risking its own capital in joint ventures with Chilean and foreign investors.

By 1986 FCh was thus directly involved in 5 enterprises in the fruit, fisheries, and livestock sector. Another batch of potential affiliated enterprises was: forestry products and wood processing as well as utilization of waste from agriculture, horticulture, and livestock farming.

SALMONES ANTARTICA

Supply/Demand Considerations

With a catch of close to 5 million tons, Chile is one of the top fishing nations in the world. By the mid-1980s fishing accounted for about 12 percent of export earnings, and was expected to be one of the most dynamic sectors of Chile's economy. Consequently FCh decided to explore the possibility of transferring fishery related technologies that would help make the sector more productive and profitable. As a result, one of the early FCh activities was preparation of a catalog of Chilean fishery resources (Bibliography, section 2.2.7). It soon became evident that southern Chile's climate and coastal geography—countless fjords, islands, archipelagoes, and protected bays—facilitates salmon production in the clear, uncontaminated and well-oxygenated water.

Indeed salmon pisciculture traces back to 1878, when introduction of this fish was first tried in Chile. Early in the 20th century, Atlantic salmon ocean ranching was successfully established, but by the late 1930s the salmon had mysteriously disappeared from Chilean waters.

FCh management decided to explore the feasibility of re-establishing salmon fisheries. Salmon is one of the delicacies appreciated by gourmet chefs around the world. Due to overfishing, natural stocks were being depleted; consequently man tried to domesticate the species. Until the 1970s the work tended to be confined to the natural habitat of salmon in the Northern Hemisphere: Alaska, Scotland, Norway, and British Columbia. There, salmon spawns in fresh waters and spends the first part of its life in its natal streams, tributaries, or lakes. A year or so after birth the salmon fry, called a smelt, migrates to the ocean. There it travels great distances feeding on pelagic crustacea and fish. During the oceanic phase, the salmon grows and matures. When ready to reproduce, salmon returns to natal streams to mate. Once the fish are spent, and a new generation conceived, parents die. During the "homing migration" to natal streams, salmon travel long distances and seek out plentiful but hard to harvest marine species. In short, salmon is a good way to convert plankton, krill, and other small marine fauna into food for human consumption.

The best salmon regions in the world have four common attributes: (a) coastlines broken up by fjords, islands, channels, and bays; (b) pure water, rich in nutrients and various forms of plankton; (c) cold lakes and rivers, crystal clear and rich in oxygen, in which salmon are born; and (d) pollution-free seas, in which the fish grow to maturity.

Salmonids are not native to the Southern Hemisphere, but Chile's Austral (Southern) Zone appeared to be a potential salmon region. This is because over more than 2000 kilometers of coastline Chile has oceanographic conditions similar to those prevailing in the Northern Hemisphere, where salmon live and pasture naturally. Near Puerto Montt, capital of Region 10, the coastline is dismembered, full of fjords, islands, archipelagoes, and bays. Most places are accessible year-round by land and/or sea. The waters are clean, unpolluted, clear, fresh, and oxygen-rich. Water temperatures and climatic conditions in the Chiloé, Aysen, and Magallanes areas are much milder than those of comparable places in the Northern Hemisphere, where salmon winter growth rates are small and living conditions severe.

In short, the readily-accessible Chilean rivers and streams should be are ideal for raising salmon in its freshwater stage. Numerous sheltered seawater areas are conducive to locating salmon farms. Population as well as industrialization trends indicate that Chilean fresh water streams and lakes would likely remain unpolluted for years to come.

Ocean Ranching Versus Cage Farming

Two technologies have been used for salmon acclimatization to Chilean waters. Initially, eggs have been hatched and the salmon raised to smelt stage in fresh water, under controlled conditions, for release and growth in a natural oceanic environment. Thereafter, the powerful homing instinct of this anadromous species forces maturing adults return to their original place of release. The mature fish is captured during or at the end of this migration. These open circuit practices are usually referred to as the "ocean ranching" technology.

The potential benefits from such ocean ranching of salmon seems enormous. Opposite the west coasts of Canada and the United States are "fishing zones" numbers 67 and 77 (so designated by the Food and Agriculture Organization of the United Nations, the FAO). In 1980, salmon landings from these areas totalled 400,000 tons. Most of these fish came from rivers in Alaska, British Columbia, Washington, and Oregon. With a coastline of about 5000 km, about 80 tons of salmon were produced per kilometer. Conservatively assuming that in southern Chile FCh could achieve production of 40 tons per linear km, the "salmon coast" of 1700 km could potentially produce 68,000 tons of ranch salmon annually. At U.S.$2 per kg, this would amount to over U.S.$200 million, or double the total value of Chile's 1982 fish and shellfish landings.

FCh had several options to choose from in adapting the ocean ranching technology.

In 1971 the Division of Fish and Wildlife of the Chilean Ministry of Agriculture (Secretaría de Recursos Naturales y Pesca, SERNAP) signed a contract with the Japan International Cooperation Agency (JICA) aimed at naturalizing Pacific salmon in southern Chile. During 1972-82 some 20.2 million eggs were imported, and 13.6 millions smelts liberated from Rio Claro, Puerto Piedra, and Bahia Ensenada Baja in Region 11. During 1979-82, 1453 returned salmon were caught at Curaco de Vélez: 1367 Chinooks and 86 Coho's.

While the feasibility of SERNAP/JICA's domesticating the Pacific salmon in Chilean waters was established, the economics of ocean ranching was far from clear: (a) as a rule of thumb it takes a return rate of about 1-1½ percent for the mature salmon to pay for raising of the smelts; and (b) the returning salmon are in the public domain, so that anybody with a fishing license can go after the salmon on their way toward the natal stream. Under those circumstances ocean ranching would evidently not be a profitable business for a private enterprise. This circumstance seems to indicate that ocean ranching should be made a publicly-subsidized activity. International agreements will settle potential conflicts about fishing rights. In contrast, private enterprise should explore the possibility of "cage" or "pen" technology, which avoids potential conflicts about property rights because the fish simply belong to their breeder.

FCh did not want to put all its eggs into one basket. The ocean ranching, or open circle technology, is slow. It is too risky to bet on a return rate of 1 or 2 percent of liberated smelts.

Thus in 1980 FCh signed a contract with the Planning Secretariat of Region 11 for the purpose of carrying out a pilot project with "cage cultivation" of rainbow trout and salmon in fresh water. The exploratory pilot cultivations produced the rates of growth shown in Table 3.3.

Table 3.3 Pilot Salmon and Trout Cultivation

Species	Origin	Weight (in grams)		Month in cages	Average gain per months (gr)
		Initial	Final		
1. Salmon					
1.1 Coho	C. Vélez	1	1.500	19	79
1.2 Keta	SERNAP	271	1.853	22	72
1.3 Chinook	C. Vélez	1	1.240	29	43
1.4 Masu	SERNAP	6	250	20	21
2. Trout					
2.1 Arcoiris	Pucón	12	2.700	15	179

The rainbow trout grew twice as fast as the Coho and Keta salmon. Yet testing was discontinued because world markets were amply supplied; low prices did not justify commercial trout production. In contrast, due to favorable demand for Coho and Keta salmon, the internal rate of return (IRR) was estimated to range from 36 percent for "farms" with annual output of 100 tons to 53 percent for output of 600 tons.

The promise of cage culture highlighted the need for substituting imported salmon eggs with locally produced eggs. In 1981, so as to gain time, FCh bought DOMSEA, an existing U.S.-owned fish ranch in Curaco de Vélez (Chiloé). There salmon fingerlings were grown to 30-40 grams in fresh water, and then liberated. DOMSEA obtained small but viable returns of salmon by the "open circuit" or "ranching" technology. DOMSEA, under FCh ownership, was renamed Salmones Antártica Ltda., made a FCh branch enterprise (empresa filial), and provided with a trademark.

The import substitution of salmon eggs was to be accomplished in 3 steps: (a) doubling the number of incubation basins from 4 to 8 so as to increase annual production of smelts from 0.6 to 1.0 million; (b) expanding new aquaculture facilities in Dalcahue (Port Chacabuco) and Rio Prat (Port Natales) in Region 12; and (c) acquiring a site in Astilleros, which would produce 2 million more smelts.

In 1983, because of promising pilot results with cage production, Salmones Antártica initiated construction of 6 pools

in Puerto Chacabuco. The purpose was to produce salmon fingerlings in fresh water to be subsequently grown out and fattened in
floating cages. The first harvest of 200 tons was projected for
1986/87, increasing to 400 tons thereafter.

In 1984 Salmones Antártica initiated construction of a
hatchery at Rio Prat aimed at producing 13 millions smelts
annually, the largest ranching operation South of Ecuador.

By 1985 Salmones Antártica thus had 4 farm sites: Puerto
Montt, Curaco de Vélez (Chiloé), Puerto Chacabuco (Aisén) and
Seno de Ultima Esperanza (Magallanes). Accumulated losses, were
expected to peak at some $1 million by 1985. The first positive
cash flow of about $100,000 was projected for 1988, to consol-
idate at about $2.5 million annually by 1990 and beyond.

Achievements

The impact of the FCh catalysis salmon farming made itself
rapidly felt. In 1985 there were 15 public and private salmon
breeders using the pen technology in 25 establishments (see
Appendix 3). Among the salmon species most commonly being used
in ocean ranching are Oncorhynchus tschawytscha (chinook),
Oncorhynchus kisutch (coho), Oncorhynchus nerka (sockeye), and
Oncorhynchus keta (chum). Most suitable for cage cultivation
have been Atlantic salmon and coho.

During 1984-85 Chile produced 500 tons of Pacific salmon, or
more than eightfold the 60 tons produced in 1981. This put Chile
third behind Japan's 9000 tons and the 1043 tons recorded for the
United States. By 1983 output from Chilean salmon farmers and
ranchers amounted to roughly 1300 tons, or 18 percent of world
production; Japan accounted for 63 percent and USA for 19
percent.

In the mid-1980s the entire FCh output went for export
markets to Argentina, Brazil, and the United States, where prices
ran from $5/kg for frozen up to $14/kg for smoked salmon. The
top price was due to the counter-seasonal harvest of Chilean
salmon, which reaches Northern Hemisphere markets in spring and
summer months, when local salmon are not available.

Post-Harvest Technology

Smoking is an age-old technology adapted by FCh in
facilities at its own pilot plant in Santiago.

To get the pink color typical of the fresh salmon, the fish
bred in pens of Salmones Antártica are being fed carotene. This
is because penned fish do not eat crustaceans, which in nature
add the pink hue to the meat. Smoking strengthens the color and
intensifies desired flavor of the meat. The main challenge is to
maintain the fish suitably refrigerated on its way from breeding

station to the smoking plant, and during subsequent storage prior to export.

Table 3.4 Output of Cultivated Salmon in Japan, U.S. and Chile, 1983

Enterprise	Tons	% of Total
1. Japan		
1.1 Nichiro	2,500	35%
1.2 Taiyo	1,000	14
1.3 Nichiro	500	7
1.4 Other	500	7
Japan subtotal	4,500	63%
2. US		
2.1 Domsea	1,260	17%
2.2 Other	143	2
US Subtotal	1,400	19%
3. Chile		
3.1 Mares Australes	600	8%
3.2 Nichiro	200	3
3.3 Salmones Antártica	250	2
3.4 Sucar	200	4
3.5 Salmosur and Others	100	1
Chile Subtotal	1,300	18%
GRAND TOTAL	7,200	100%

Source: Pacific Fishing, February 1985.

The economies of scale in post-harvest treatment are such that FCh helped set up an Association of Salmon Breeders and Exporters (ASBE), aimed to jointly own and operate smoking and storage facilities as well as subject their product to strict quality control.

FCh also undertook pilot trials with irradiation of smoked salmon. Although the results have been promising, the technology has not yet caught on. (A report detailing this work can be obtained upon request from the author.)

Roads Ahead

On basis of data at hand it is possible to speculate about the type of enterprise that might lead more potential producers to might want to enter commercial salmon production in years to come. Work is already underway aimed at transferring or adapting technologies from livestock farming to salmon ranching. In the first place Canada's Department of Fisheries and Oceans Laboratory in Vancouver is injecting salmon with chicken and bovine growth hormones produced by Anigen (Thousand Oaks, California). The treatment seeks to provide a cost-effective way to accelerate growth and make more protein-rich fish available on a worldwide basis.

In 1984 initial research was done in Chile to ascertain the feasibility of moving salmon farming to the northern, arid desert areas of Chile. The work was catalyzed by Peter Brown, a former FCh staff member. Brown has been trying to persuade fish meal manufacturers to look into the feasibility of using part of the rich pelagic fishery of about one million tons annually as basis for inexpensive salmon feed.

In June 1984 Pesquera Playa Blanca, one of the fish meal companies in Northern Chile, obtained 500 pacific chinook smelts from FCh and brought them 2250 km north to Caldera. The smelts weighed 45-50 grams and were conditioned to salt water before transport in large plastic bags, filled with seawater and placed in styrofoam boxes. Oxygen was provide during the 15-hour flight. After 10 days, some 350 fish were in prime condition and mortality stopped.

The smelts were fed fresh jack mackerel or pilchards, which were ground and mixed with fishmeal. Water temperature over the years has not gone above 21°C. Growth was spectacular. By November 1984 the average smelts weighed 333 grams or almost sevenfold their June 1984 weight. At least one growth cycle would be needed to provide some indication of maturation rate.

The challenge of supplying fresh water in the arid northern desert seems on the way to being resolved. In October 1984 some 14,000 Coho smelts were trucked from the southern bay of Calbuco to Tongoy, some 1560 km to the north. Only 17 fish died in transit. It appears that acclimatization is possible using recirculation and progressive addition of salt water. This "southern salmon technology", pioneered by FCh, can likely be further adapted for transfer of young fish to northern Chile, where there is plenty of inexpensive feed and the cold Humboldt current keeps water temperatures at suitable levels.

In short, Chile was the first to introduce Pacific salmon to the Southern Hemisphere. Both Australia and New Zealand have since achieved this feat. Yet Chile remains the only country to have succeeded in both caged cultivation and open-sea ranching of coho salmon.

Would these developments have taken place without FCh involvement? In 1980, when FCh started its salmon domestication work, only two small entrepreneurs were active. By resolution 633 of December 31, 1984, the Subsecretariat of Fisheries authorized Salmones Antártica to install and operate a cage salmon cultivation establishment in Chanquitad on the Quinchao island of Region 10. By 1985 some 14 companies were farming salmon and producing 800-1000 tons, mostly for export.

According to Anthony Wylie, Director General of FCh, "even without Fundación's research and financing, no doubt someone would have spotted the potential. But he would have had trouble getting contacts and the right technology. The boom in salmon production would probably have happened anyway, but at a later date [As things looked in early 1987, Chile is likely to] become a big producer of salmon in the near future... production [of] 8000 to 10,000 tons of salmon by 1990/92 seem feasible. Early comers... will certainly choose best available places [and]... pave the way—hopefully at a reasonable cost—for those commencing production later."

These projections were repeated during the March 17-19, 1987, FCh-hosted International Outlook Seminar for Salmon Culture in Chile. The participants agreed on the following projections to 1990/92 for world production of cultivated salmon (in 1000 tons), shown in Table 3.5.

Table 3.5 World Production of Cultivated Salmon, 1986 and Projections to 1990/92

Country	1986	1990/92
1. Norway	45.5	100.0-120.0
2. Scotland	10.0	20.0
3. Ireland	1.0	10.0
4. Faroe Islands (Denmark)	0.6	10.0
5. Iceland	0.2	10.0
6. Chile	2.0	8.0-10.0
7. World Total	59.3	158.0-190.0
8. Chile as % of World Total	3.8%	4.2-5.6%

Source: Clara Munita O., Caution and Courage about Salmon Culture in Chile, Chile Pesquero, May 1987, p.45

In short, with Chilean production of 8000 to 10,000 tons by 1990/92 (some participants expected it to be 12,000 to 15,000 tons), Chile's share of world output of farmed salmon would increase from 3.8 percent in 1986 to somewhere between 4.2 to 5.6 percent. This appears a reasonable estimate.

Achievement of those goals will face some challenges that must be solved now:
(a) there will be pressure from (fish) farmers in the Andean Free Trade Association (AFTA) to restrict imports of salmon if there is a downward trend in prices;
(b) there is lack or adequate cold storage, processing plants, and reliable feed supplies;
(c) Chile does not have enough qualified personnel to prepare and execute salmon farming projects;

(d) there are no facilities at Chiloé to test for diseases and certify quality;
(e) exports of 10,000 tons of fresh or smoked salmon would require 220 charter flights; and
(f) Chile needs to be self-sufficient in salmon egg production so as to prevent importations of diseases as well as for economic reasons.

Furthermore, the Salmon and Trout Producers Association, which was founded in 1986, urged initiation of a promotion program aimed at increasing demand for Chilean fish. (In 1986 Scotland invested $0.3 million in promotion, and boosted its 1987 return sixfold, to $2.5 million.)

The United States would likely be the first target for such a promotional effort, because it is the main importer of fresh salmon. In 1986 it imported close to 13000 tons, with a FAS value of $78 million, or $6/kg. The following data show that unit prices varied all the way from a low of $2.62/kg for Canadian salmon to a high of $7.90/kg for Scottish salmon:

Table 3.6 Unit Prices of Salmon from Major Countries of Origin, 1986

Country of Origin	Volume (tons)	Value (US$ million)	Value (U.S.$/kg)
1. Norway	8,860	62.1	7.00
2. Canada	2,482	6.5	2.62
3. Chile	681	2.9	4.28
4. United Kingdom (Scotland)	367	2.9	7.90
5. New Zealand	161	1.0	6.29
6. Netherlands	105	0.8	7.50
7. Others	271	1.6	6.01
Total	12,931	$77.8	$6.02
Frozen salmon	5,501	–	–
Chile as % of total	5.3%	3.7%	71.1%

Source: Clara Munita O.

The consensus of the 150 participants in the FCh-sponsored March 1987 pilot seminar was that "a good job was done."

CULTIMAR OYSTER FARMING

Supply/Demand Considerations

Similarly to salmon, oysters are a much sought after delicacy in North American and West European markets.

Artificial breeding of oysters is a technology started by the French emperor, Napoleon III, when the popularity of the dish made oysters an endangered species. In Southern Chile oysters have been produced for many years. Native species grow slowly, and geographic distribution is limited. Seed production was erratic. Under those circumstances it seemed justified for FCh to introduce a new species, the Pacific oyster Crassostrea gigas.

FCh experts, working with foreign consultants, planned the importation and rearing of Pacific oysters, while suitably adapting Japanese technology to Chilean circumstances. Local culture was established in Coquimbo with initial participation of the Universidad del Norte.

There was one major problem. The cold ocean water prevented the oyster from reproducing in Tongoy. The continued seed importation was considered risky because the danger of introducing foreign diseases lethal to fauna in Chilean waters. In 1981 FCh therefore decided to set up its own oyster hatchery.

Selection of Species

During 1978, in the wake of testing 8 different shellfish species, Pacific oyster seeds were brought in from California. Results exceeded expectations: spawning was induced on a year-round basis and production could be controlled. In the Coquimbo facilities the Pacific oysters matured in 9 months. In contrast, native oysters require 3-4 years to attain market size.

In 1981, based on the favorable pilot production and market tests, FCh initiated construction at Tongoy of a laboratory and oyster farm with an eventual capacity to produce 44 million oysters for local consumption. FCh directors felt that afavorable export market outlook, the increase of local seafood supply, and the contribution to employment justified direct investment in the venture. Consequently, Cultivos Marinos Tongoy (CULTIMAR) was set up. FCh retained a majority of the shares. In early 1983 grow-out was initiated in a hatchery.

Achievements

By October 1983—or a mere 6 years after initial introduction of the Pacific oyster into Chilean waters—the product was introduced into national and international markets. FCh presented oysters to buyers via the International Food Fair (ANUGA), held in Cologne (West Germany), and the International

Seafood Conference, held in Vienna (Austria). On both occasions
the product was favorably received. In August 1984, CULTIMAR
started to deliver large juvenile oysters to private grow out
centers throughout Chile.

The outlook was so bright that by mid-1984 contract oyster
farming was initiated in Region 10. Some 40 low-income families
obtained 1000 seeds for grow-out. By year-end some favorable
results were at hand: mortality was less than 1 percent, or
substantially below expectations. Based on 320,000 seeds (8000
to each of 40 families), to be distributed monthly, output of
280,000 was to be reached by 1987/88.

Similar to relationships between egg hatcheries and chicken
farmers, FCh helped private fish farmers get into business by
preparing feasibility studies, building facilities, and training
personnel. The time lag between installation of a cage farm and
marketing of first crop was reduced to about 1½ years.

By 1985, or a mere 6 years after FCh brought the Pacific
oyster from Japan, the Tongoy oyster, named after the location of
the FCh breeding station, has been firmly established. The FCh
oyster is different from native Chiloé island oysters: it is
bigger, has a deeper shell, contains more meat and juice, has a
mild taste, grows faster, and is available year-round. Gradually
Chilean housewives became accustomed and appreciative of its
qualities, and learned to prepare suitable dishes. The final
seal of approval came in 1985 when Hernan Eyzaguirre, the Chilean
gourmet/columnist, publicly confessed his conversion after
tasting a FCh oyster "quickly cooked in champagne and garnished
with vegetables." Eyzaguirre then applauded FCh for supplying
the local market and opening up possibilities for exports.

The oyster seed can survive out of water for up to 72 hours.
This makes it feasible to undertake shipments to the Northern
Hemisphere during the November March season. The demand in the
United States and Mexico amounts to at least 15 million seeds.
In 1985 FCh made a pilot air shipment of oyster seeds from its
CULTIMAR facility to Mexico. The 1 million seeds, worth $6
million, were successfully shipped in containers held at 4-6°C.
In short, by 1986 Chile was on its way to becoming the world's
leading producer of off-season oyster seeds.

Roads Ahead

The oyster breeding park in Tongoy is an installation with
built-in flexibility. It is adaptable to other marine species
that could be farmed in Chilean waters, for example, turbot
(Scopthalmus maximus), a European flounder of high commercial
value. Significant advances in adapting turbot to marine
conditions on the Chilean coast have already been made. Chile
could therefore become the first country in South America to
successfully cultivate this valuable saltwater fish.

Extra tanks and water pumping equipment make it possible to also raise scallops and abalones at Tongoy.

Another promising species is ostion, which is in great demand and well adapted to the natural banks in the vicinity of Tongoy and Guanagueros. The Japanese government signed a contract with the University of the North for producing ostion seeds. The Japanese contemplate a total investment of $5 million.

CAPRILAC GOAT CHEESE

In comparison with the imaginative oyster farming and salmon ranching ventures, the goat ranching project was conceived along more pedestrian lines.

In 1982 the Corporation for Social Development in Rural Areas (Corporación para el Desarrollo Social del Sector Rural, CODESSER) initiated an improvement program for production of goat milk and cheese. FCh served as a broker by bringing into Chile technology readily available in France. This was done by special agreement of FCh with the French Institut de Téchnique de l'Evelage Ovin et Caprin (ITOVIC), one of the leading European research and development centers for goat and sheep production. ITOVIC work culminated in 1983 by creation of the Comercializadora de Productos Caprinos (CAPRILAC), at the Agricultural School in the Ovalle area of Region 4. CAPRILAC was a joint venture in which FCh had 70 percent of stock, CODESSER 30 percent.

The plant was designed initially for processing of 3000 liters of milk daily, to be expanded to 6000 liters if required. The investment was about $70,000, with an anticipated 3-year pay-out period.

By early 1984 CAPRILAC cream cheese and herb-flavored cheeses were available in Jumbo and Almac supermarkets in Santiago. By late 1984 the first goat cheese were ready for export. Due to shortage of raw milk during the winter months, production of cheese lagged behind market demand. The inability of goat herd owners to boost milk production adequately became a financial burden on CAPRILAC. The cheese plant was closed.

PROCARNE

Supply/Demand Considerations

In 1980 some 164,000 tons of carcass beef was wholesaled in Chile. Over half the volume was accounted for by the Santiago Metropolitan Area (SMA), where channels of distribution were as shown in Table 3.7 (in 1000 tons):

Table 3.7 Carcass Beef Wholesaled in Chile, 1980

Type of Outlet	Amount	% of total
1. Butcher shops	62.3	69%
2. Institutions	16.5	18
3. Supermarkets	11.4	13
Santiago sub-total	90.2	100%
Provinces	77.5	
Total Chile	167.7	
Santiago as % of Total	54%	

Source: Juan Pablo Torrealba, Prefactibilidad de Procesamiento y Distribución de Carne en Caja (Pre-feasibility of Processing and Distribution of Boxed Beef), FCh, May 18, 1983, p. 11.

Self-service supermarkets tend to be located in middle- and high-income neighborhoods of Santiago; they require meat of quality superior to what normally is being provided. The traditional trade tends to sell carcasss meat in inconvenient, unsanitary and wasteful ways, a situation similar to the United States during the 1960s. In the early 1970s the vacuum-packed boxed beef technology was introduced in the United States and took off like wildfire. By 1985 over 85 percent of all U.S.-produced red meat was being wholesaled in that form.

Vacuum Packing Technology

Under those circumstances FCh saw an opportunity for transferring to Chile the vacuum packing technology, with standard meat cuts sent to market in cartons as "boxed beef." The associated benefits of this processing and marketing technology were to be as follows:
(a) slaughter of cattle in rural production areas, rather than at the terminal facility in Santiago, reducing cattle bruising and losses incurred in transport;
(b) removing a source of urban environmental pollution caused by the presence of slaughter plants in or near cities;
(c) sanitary vacuum-packaged and refrigerated meat can maintain its quality (shelf life) at least 4 times longer than traditional carcasss meat;
(d) cardboard cartons of boxed beef, weighing about 30 kg, are easy to handle;
(e) buyers gain flexibility in efficiently cutting boxed meat to individual specifications; and

(f) carefully graded boxed beef is priced according to quality, thus rewarding livestock breeders for not selling live cattle on basis of weight only, a traditional method that has prevailed in Chile for a long time.

Boxed beef has not been introduced to Chile previously because the technology apparently calls for radical changes in methods of livestock procurement, slaughter, and processing, as well as distribution. To do that requires overcoming existing vested interests at each stage of the marketing channel.

Under those circumstances FCh saw an attractive opportunity for providing a demonstration of overall feasibility of such a venture, based on a plant with daily capacity of processing 125 head of cattle, or 30,000 head a year.

The cash-flow analysis indicated losses for 1983, 1984, and 1985, with stabilization and positive results in 1986 and beyond. The expected internal rate of return (IRR) was estimated to be about 15.2 percent; with increase of sales by 10 percent, IRR would be 31.8 percent, with a boost of 20 percent, IRR would be 46.5 percent.

Establishment of a Vertically-Integrated Industry

Following U.S. experience with the Village Meats and Cryovac corporations, Donald W. Long came to Chile in 1978 to help with the initial pre-feasibility study. Preparation of blueprints started in July 1983, construction was initiated in December 1983, and operations went on stream by July 1984. The plant was designed for 125-140 head/day and started to operate with 64 head/day, or little below half the target capacity. There is sufficient space on the 30-acre site to expand the plant. Long stayed on as general manager.

The PROCARNE plant uses state-of-the-art technology. The building has a steel-panel shell, with prefabricated panels that insulate refrigerated areas. The rail support structure is free standing and independent of building support. The plant has its own water well and its own effluent treatment facility. Compared with fuel oil, the wood-fired steam boiler reduces energy costs by about 40 percent. Much of the equipment was supplied by Omeco-Boss of Omaha, Nebraska. Some parts were fabricated locally.

The packing plant is located at Osorno (Region 10), in the center of a major cattle producing area. To assure adequate supply of live animals, FCh established a joint venture with the local cattle breeders association.

Commission buyers procure animals. The cattle are then killed at a nearby slaughter house, which had to substantially upgrade its facilities and operation so as to meet PROCARNE's high sanitation standards. Carcass quarters are held overnight at the slaughter plant and then shipped about 3 miles to

PROCARNE. There they are kept an additional 12 hours in a 32°F cooler prior to cutting and boxing.

The plant is designed in a circle. Shipping and receiving is done from a common dock. Once quarters are received, they move into a cooler and are hung on rails, where some pre-breaking occurs. Processing begins on custom-designed on-rail and table-boning equipment, followed by bagging, vacuum packaging, shrinking, and boxing. The processing room is held at 38°F. The boxed beef is palletized and moved to a 28°F cooler for twice weekly dispatch to Santiago. The shipping is done in mixed loads on trucks used by Felco, a local sausage processor. Short ribs for stews are frozen and sold locally.

Cuts are similar to those produced in the United States. PROCARNE is using the outside lean of the plate to cover the bone areas on ribs. This saves the cost of the bone guard, which can be sold separately; there is a ready market for that cut of lean meat. Packaging materials have to be imported and are relatively expensive.

A private grading system has been developed, and is being enforced by a strict quality control program. The system consists of 5 grades, based primarily on maturity, extent of marbling, and fat cover. The two top grades (Premium and Super) are boxed in white boxes; the three lower grades (Select, Standard, and Industrial) are packed in manila-colored boxes. FCh is assisting the government to translate the PROCARNE specifications into an official grading system.

All initial employees of PROCARNE were new to meat packing, and had to get intensive on-the-job training. To ensure order and cleanliness, the locker room is maintained by teams of employees, with assignments posted on bulletin boards.

Workers operate in 9-hour shifts for a 5.5-day week. They are being paid monthly by check. During winter it gets cold in southern Chile. Because many workers either did not or could not afford to bring an adequate lunch, they became hungry by mid afternoon. Productivity started to fall off about 3 p.m. PROCARNE began supplying lunches through a contract caterer. The increase in output during late afternoon more than pays for the meals.

Initially the PROCARNE branded meat was sold to supermarkets, the wholesale trade, and institutions (hotels, hospitals, restaurants, schools, military installations) in the nation's capital. Once established in that market PROCARNE plans to expand its plant as well as distribution network.

Achievements

By 1985 it was evident that PROCARNE boxed beef, which then barely represented 5 percent of total red meat sales, was to be the "tail that would wag the dog"; that is, the demonstration effect caused a gradual restructuring and modernization of the entire subsector. This is precisely the type of catalytic effect that FCh strives to achieve.

The PROCARNE plant serves as example of how to adapt U.S. technology to the needs of a country with an incipient meat packing industry. Changes in physical handling of meat (hardware) were used as catalyst for changes in marketing as well as in plant management (software).

The financial performance has not quite lived up to expectations. FCh management expected cashflow to turn positive by about 1987.

BERRIES LA UNION (BLU)

Supply/Demand Considerations

A large number of berries grow wild in Chile: black-, goose-, blue-, cran-, rasp-, and strawberry. Because of their perishability, harvesting for commercial sale has so far been difficult. Small plantations of straw- and raspberries started to be established only in the late 1960s.

In 1978 the first exportation of fresh raspberries took place. It was soon followed by occasional exports of strawberries to the United States, Canada, and Western Europe. In the early 1980s, due to poor quality of the fruit, major rejections of Chilean berries at ports of entry were increasingly occurring. Still, by 1983 exports of berries in fresh and processed form reached 5600 tons with an FOB value of $6.7 million, or $1200/ton.

Preliminary studies, undertaken by FCh, indicated Chile enjoyed a comparative advantage in terms of growing season, soils, climate, and availability of labor. Mid-southern Chile has favorable climatic conditions, well drained soils, and ample irrigation water. The summers are relatively cool. This greatly facilitates post-harvest handling of the berries. There are no frosts. A large bee population pollinates the flowers. There is abundant labor capable and willing to work. With the exception of biannual strawberries, most other berries bear fully at 3-4 years. The bushes have a commercial life of 15 years or more.

Technology Selection and Diffusion

In 1984 the Fruit and Vegetable Division of FCh initiated a Technical Assistance Program for Berry Products (PROBERRIES), aimed at transferring U.S. and European technologies for growing, harvesting, packing, and marketing export-quality berries. The extension service, offered to individual farmers by PROBERRIES, was designed to help to: (a) identify agro-ecological attributes of individual farms and select suitable varieties; (b) supervise planting; (c) monitor for disease and insect infestation; (d) instruct in use of herbicide, fertilizer, and plant protection measures; (e) establish harvesting, grading, packing, and refrigeration procedures; and (f) inspect and certify quality of packed berries prior to shipment.

Achievements

The initial FCh sponsored research and promotional effort has been so successful that in November 1984 a group of producers and entrepreneurs in La Union, Rio Negro, and Osorno formed a committee for creation of a regional corporation for production and export marketing of berries. The initial issue—300 shares at 250,000 pesos each (about $2000)— was to be placed in three batches: 40 percent taken up in December 1984, 40 percent in June 1985, and 20 percent in December 1985. FCh was to provide: (a) seed capital of about 20 percent of the total initial investment, estimated at about $0.7 million, and (b) technical assistance needed for production grading, packing and export marketing of the berries. In 1985 some 35 hectares were planted by members of Berries La Union (BLU).

Before inauguration of the BLU facilities, FCh organized a U.S. study tour for Plutarco Dinamarca, manager of BLU. He went to the Universities of California (Davis), Oregon (Corvallis), Minnesota (St. Paul), and the U.S. Department of Agriculture (Beltsville, MD).

BLU built a refrigerated warehouse and packing facilities for handling its own production as well as for performing marketing functions for other producers in the region. It is expected that about 60 percent of the fruit will be exported in fresh form; surpluses would be frozen and/or sold fresh in Chile.

Because only first-class berries can be exported, FCh initiated a corresponding quality certification program. BLU is to have a positive cash flow by about 1988.

FORESTRY RELATED ACTIVITIES

Background

The Chilean forestry sector is looming as a giant, gearing up for exploitation of the huge forest tracts that will reach maturity in Chile's central-south region during the 1990s.

Chile has one of the world's largest areas of planted forests , and already ranks first in the radiata pine species. As a result of a vigorous government development program, coupled with subsidies for new plantations, over one million hectares of radiata have been added to existing plantations. The subsidy is expected to continue through 1994, when new plantations are projected to exceed 1½ million hectares.

Of the countries with important radiata pine forests, only Chile has real potential for further expansion; it could double the existing radiata planted area. In the United States radiata pine (a native California species), normally matures in 30-40 years; in Chile only 18-20 years are required. Because a large percentage of the new forests were planted ten years ago, these vast tracts of woodland will become ready for exploitation in 5-10 years. These huge resources are expected to produce between 20 and 40 million cubic meters (m^3) of wood. Chile does not have the infrastructure of plant and equipment to process the growing quantity of wood. The forestry sector will require new capital investments exceeding $1 billion by 1990, and totaling between $2.5 and $3.0 billion before the turn of the century, to process the new resources, including:

(a) 12 new sawmills, each of 100,000 metric ton/year capacity, with kiln-drying and planing equipment;

(b) 4 plants to process an additional 20,000 tons of particleboard and 140,000 tons of fiberboard;

(c) construction of 10 plants of 10,000 m/yr capacity, plus expansion of existing installations, for production of integrated, export-oriented, furniture and housing components;

(d) 4 plants to process an additional 1.2 million tons of Kraft pulp;

(e) 4 plants to produce an additional 260,000 tons of paper of various types; and

(f) construction of a bulk ship loading facility to handle wood chip export.

During 1985 and 1986, recognizing the possibilities for investment in the Chilean forestry sector, foreign investors have closely scrutinized available opportunities. New Zealand and Saudi interests have already invested several hundred million dollars, both for equity in existing plants and in purchasing forest tracts.

Supply/Demand Considerations

Since about 1980 much of FCh's promotional effort has focused on the forestry sector. The work has only gradually gotten underway. This is because it looked like Chile was "of course" already deeply involved in major exploitations of the huge natural resource by means of a well developed pulp and paper industry. Forestry work was thus tucked away, with "micro-processors," in a corner of the FCh organization chart.

In the early 1980s Dr. Roberto Echeverría was enticed by FCh to return to Chile from a position as economist at the World Bank—a case of "brain gain." His first task, as head of planning and programming, was to design methodology guidelines for systematic identification, selection, preparation, and full-fledged feasibility appraisal of projects. He soon became impressed with the enormous development possibilities in the huge forestry resources, which so far seemed to have been subjected to

benign neglect. There appeared to be no end to potentially
attractive investment opportunities. Before zeroing in on
specifics, a thorough sector study was needed to help identify
problems, sort out ideas, assign priorities, prepare projects,
and start looking for investors within as well as outside of
Chile. The study, undertaken in 1982/83, more than confirmed Dr.
Echeverría hypotheses. By mid-1995 annual growth of radiata
pine in existing plantations would quadruple. Radiata pine wood
is a remarkably versatile material, low in cost and, when
properly handled, most suitable for a wide variety of
applications. FCh concluded that if Chile failed to gear up for
using this enormous resource, it stood to lose it.

Under those circumstances FCh decided to set up a Forestry
Department under Dr. Echeverría's leadership. During 1982 and
1983, ten projects were defined, studied, and evaluated. While
very diverse areas were covered, all called for transfer of
advanced technology: radiata pine nurseries; construction of a
high technology sawmill in Region 8; creation of a silvichemical
industry based on tar oil (black liquor) from pulp and paper
mills; fabrication of radiata pine plywood panels; construction
of Energy Thermic Housing (ETH); certification of quality of
structural panels; use of forestry residues; production of
elaborated wood components, and so on.

There seemed no end to that list of important things to do.
By 1984 priority was assigned to 5 areas: (a) application of
small-scale technologies for production of furniture, tools,
toys, doors, and windows; (b) use of residues from forests,
sawmills, and paper and pulp factories to provide fuel for
co-generation of electricity; (c) new uses for resins and
tanning; (d) installation of a decorative plywood mill in Chiloé,
based on use of native woods, and on radiata pine in Coelemu; and
(e) promotion of mass production of Energy Thermic Housing (ETH).

Energy Thermic Housing (ETH)

The case of ETH illustrates how a major program got
underway. With over one million hectares (2.6 million acres) of
pine plantations, located mostly in Region 8, Chile has abundant
and rapidly growing forest resources. FCh saw a particularly
striking opportunity for technology adaptation in use of wood for
housing. It is one of the paradoxes of Chile that, in spite of
abundant wood supply, most family houses are made of masonry:
bricks, cement, adobe. In contrast, Northern Hemisphere
countries—the United States, Canada, and Scandinavia— use
construction technologies based on intensive use of wood.

A first attempt at entering the field was made during 1976,
when Chile exported some 6000 prefabricated houses to Venezuela.
These exports turned out to be a "flash in the pan" simply
because the Chilean products were neither competitive in price
nor in quality. In early 1982 Pro-Chile started to retest the

waters. A committee on prefabricated housing was set up with participation of 11 national wood processing enterprises. The objective was to open up export markets in the United States, the Caribbean, Central America, the Middle East, and North Africa, and to increase exports from $5 million in 1985 to over $30 million by 1990.

It soon became evident that to overcome the large transport costs to distant markets Chile had to offer a prefabricated house with unique advantages. However the technology for such a product was not at hand and therefore would have to be brought in.

Consequently, FCh launched an inter-institutional consortium with the University of Chile (UC/FCh) to promote intensive use of wood in construction in Energy Thermic Housing (ETH), using structural wood, with interior and exterior walls and the roof base made of plywood. Concrete is used only for the foundations. In order to give ETH houses a more traditional appearance, interior and exterior walls are finished with a masonry-like surface.

In comparison with masonry, ETH systems offer at least nine specific advantages: (a) less cost, due to better utilization of materials, an increase in useful surface area, and 5-10 percent in additional interior space; (b) insulation, provided by ETH, saves up to 70 percent in heating energy; (c) with aid of critical-path programming, ETH construction time can be cut by one-third; (d) ETH resists structural damage caused by earthquakes; (e) vapor barriers in ETH protect walls from destructive moisture; (f) due to ease with which ETH walls are assembled and disassembled, costs of adding to buildings are low; (g) ETH construction work is cleaner and quicker than masonry; (h) repair of water, light, plumbing, and heating systems in ETH is faster and less expensive; and (i) ETH construction is fire resistant.

Consequently, the inter-institutional UC/FCh consortium concluded that ETH is a technically valid alternative for reducing Chile's mounting housing deficits.

FCh's marriage brokerage proceeded apace, arranging for joint ventures with foreign partners who have the know-how for making and marketing prefabricated houses. By the end of 1984 at least two joint ventures were established between Chilean enterprises and foreign partners: National Homes of Canada (NHC) and Oregon Transpacific Housing (OTH), which provide technology. By mid-1985 two Scandinavian companies seriously nibbled at the bait: Kahrs Maskiner from Sweden and Rauma Repola from Finland. A site for the first plant was being sought near a port in the Talcahuano Concepción area.

By 1985 FCh had enough confidence in the validity of ETH that it launched a program of quality certification. In contrast to the FCh quality seals for food products, this one is being issued jointly by UC/FCh.

On Sunday, March 3, 1985, disaster struck. An earthquake measuring 8.00 degrees on the Richter scale dramatically demonstrated the vulnerability of the "hard and solid" masonry building materials. The property damage, inflicted by the disaster, was awesome: one million people suffered damage to their dwellings; tens of thousands of homes were ruined beyond repair; several new high-rise condominiums in the fashionable resorts of Viña del Mar were destroyed; the Light House apartment building of Reñaca was tilted like the Tower of Pisa, and had to be torn down; colonial and traditional adobe-style buildings in historic Old Santiago were damaged beyond repair. Chile's housing deficit jumped to 750,000 homes. The earthquake struck during the day when many people were not in their homes. Casualties were therefore "only" 177 dead and 2500 injured.

Destruction to the important ports of Valparaíso and San Antonio—main gateways to Chile's foreign trade—halted overseas shipments of fruit and copper. A number of highways and bridges were destroyed. Total damage was assessed to $2.0 billion.

The country made a fast, determined, and systematic recovery. By mid-1985 ports had restored loading and unloading facilities sufficiently to make them operational, so that fruit shipments could reach Northern Hemisphere markets at peak seasonal prices.

A number of government agencies participated in helping repair and build emergency dwellings, supply services, arrange for deferral of payments, and grant special loans and credit facilities. Chile received emergency help from at least 32 nations and multinational organizations, including credit facilities agreed on with the Inter-American Development Bank (IDB) and the World Bank.

Wooden emergency shacks and houses were erected in quake affected areas. As is usually the case, these emergency structures are likely to be around for a long time to come.

FCh headquarters were also hard hit by the quake. While structural damage to the reinforced concrete office, laboratory, and pilot plant buildings were being repaired, FCh evacuated its offices to three rented locations in the Providence sector of Santiago. By Thanksgiving 1985 the staff was able to move back into its offices. The laboratory moved to the nearby Instituto Tecnológico (INTEC). The pilot plant continued to function on a limited basis. The damage to the buildings, which house the laboratories, was more serious. It took until early 1986 to make FCh fully operational. In short, FCh's preaching for wood-based ETH construction technology, while building masonry structures for its own use, is one of those "do as I tell you, don't do as I do" counseling exercises.

The earthquake gave FCh an opportunity to practice seren-dipity in technology transfer. FCh tried to use the earthquake aftermath to systematically accelerate promotion of ETH. By mid-1985 it brought a U.S. forestry mission to Chile to help identify commercial opportunities for wood in constructing houses. Specifically the mission aimed to help intensify use of wood in house construction, establish contacts with Chilean foresters and timber processors, and increase trade between the United States and Chile.

Led by Congressman Don Bonker (D. WA), the mission consisted of senior executives from enterprises in forestry, house building, and related sectors together; specialists in economics as well as staff of the International Trade Administration of the U.S. Department of Commerce (ITA/USDC); and an officer of the American Plywood Association (APA), a trade group that represents about 75 percent of U.S. producers.

The mission suggested that an ETH demonstration project be set up as initial step in a long-term promotion campaign, and committed themselves to donate materials for construction of ten buildings.

A detailed pre-feasibility study of ETH was made for units with 45 m² of floor space, using the ETH system and roof structures designed so as to enable addition of a second story, if and when needed. The UC/FCh consortium prepared terms of reference that guided the pre-feasibility study by Sergio Bezanilla Ferres, Consulting Engineer. The UC/FCh consortium then invited bids from construction companies interested in putting up some of the pilot homes in Central Santiago. The demonstration houses—including the U.S.-donated structural plywood panels, uprights, crossbeams, doors, windows, and frames—were erected by local builders in the Las Rejas suburb of Santiago.

Roberto Busel, Vice President of Busel & Busel, a leading Chilean construction firm, was invited by the National Association of Home Builders (NAHB) in Washington, D.C., to a study tour of the United States. There he obtained guidance in planning construction as well as in methods for training builders, foremen, and workers not acquainted with ETH technology. The Corporación de Capacitación de la Cámara Chilena de Construcción (CCCCC), the Chilean counterpart to NAHB, participated in this outreach effort.

Busel & Busel constructed one of the ETH demonstration units in the new residential housing subdivision on a parcel of flat land about 3 blocks below FCh headquarters. There visitors could see for themselves how an ETH model house looks, and how it has been furnished by Ana Maria Klammer Borgoño, one of Chile's leading decorators. Busel & Busel was also chosen to supervise the work of two other contractors (Empresa Constructora Moller y Pérez and Cotapos and Habitec), who won bids for construction of the ten ETH demonstration units.

In short, Roberto Busel helped pave the way for making ETH routinely acceptable under local housing ordinances and codes by: (a) introducing UC/FCh quality control into the Chilean ETH construction industry; (b) simplifying processing of construction applications for ETH units; (c) facilitating access to long-term financing by potential buyers; and (d) removing myriads of other hurdles that might stand in the way of rapid spread of ETH technology.

The model ETH units were being sold to buyers qualifying for the 400 development units (unidades de fomento) housing subsidy, which in late 1985 amounted to about $6400. The total cost put the model ETH houses in the upper-middle class bracket, selling at prices about equal to traditional masonry structures. Because ETH is a nontraditional technology the initial resale value of the houses was expected to remain relatively low. The advantages of energy efficiency, soundproofing, and fire and moisture resistance must be traded off against this disadvantage, a risk run by any innovator.

While the demonstration ETH houses were being constructed, UC/FCh was exploring ways to mass produce ETH, so as to make it initially affordable by middle- and lower-middle class families. Furthermore, the ETH pilot project, initially confined to the Santiago Metropolitan Area (SMA), was to be subsequently adapted to requirements of southern Chile. In short, the powerful March 3, 1985, Chilean earthquake was a brutal laboratory test of building technology. It yielded important lessons for Chilean architects, designers, and city planners.

There is, of course, nothing new under the sun. Relatively "light" masonry and cement became fashionable in the mid-1930s when it started to replace traditional wooden houses with stucco finish and heavy adobe filling between the walls. The modern ETH technology combines the best elements of masonry and wood. That is why, since the early 1950s, it has become a prevalent building method in the United States and Canada. All that FCh aimed to do was to shorten the time lag needed for adaptation of the technology to Chilean circumstances.

Expansion of ETH would require reliable supply of quality raw materials. FCh is helping provide that back-up by supporting a bevy of related programs: improvement of radiata pine nurseries and plantations, increasing productivity in saw milling, and materials research and testing.

By early 1987 the ETH technology was hovering above the launching pad, with thumbs up for takeoff. In spite of the earthquake, which theoretically should have been a powerful sales promotion tool, the going remained surprisingly tough. Only by the end of the 1980s will we know whether ETH has gone into successful orbit.

Improvement of Radiata Pine Nurseries

Most of the early radiata pine plantations were established without application of suitable technology. Many existing tree farms therefore struggle against weed invasion, nutrition deficiencies, drought, and shortage of good nursery stock for replanting.

FCh is lending a hand with adapting technology used in New Zealand, a Southern Hemisphere country with ecological conditions similar to that of Chile. In 1985 FCh brought to Chile a New

Zealand forestry technician to help train the staff of eleven pine nurseries, and to widely disseminate the acquired knowledge among enterprises buying trees for establishing new tree farms as well as for improving existing plantations.

Saw Milling Center

In mid-1985 a $350,000 technological center for saw-milling was established at the University of Concepción (UC). Its purpose was to help improve productivity of sawmills in Region 8, where most pine plantations are concentrated. Ing. Gustavo Chiang, who previously worked for FCh, was selected as head of the center. Initial focus was on in-service training of workers in maintaining saws for milling boards and knives for peeling plywood logs, as well as adjusting equipment to individual logs so as to increase yield of quality products, and reduce sawdust and waste.

If cuts could be narrowed from 10 to 7 millimeters, sawdust losses could be reduced by 30 percent. A plant milling 100,000 m^3 annually—a common size in Region 8—could thus boost output by 3000 m^3, adding $150,000 to revenue. If merely 3 plants in Region 8 were to improve sawmilling, by using proven technology, the initial $350,000 investment in the sawmilling center could be paid off in one year.

FCh suggested that sawmills in Region 8 become partners of the UC/FCh Center, while FCh would continue to provide technological support. After a test period of 3-5 years the mills could then decide whether they wanted to become complete owners of the facility, while buying consulting services from FCh, if required.

Materials Testing

In 1985 the UC/FCh consortium helped establish a Materials Research and Testing Institute at the University of Chile (Instituto de Investigaciones y Ensayos de Materiales, Universidad de Chile, IDIEM/UC). On January 28, 1985, a Laboratory for Panels and Adhesives was inaugurated in the Department of Wood Technology at the University of Chile. This was a culmination of human resource development activities initiated in 1981, aimed at training abroad of academic staff to teach industrial extension and research.

Utilization of Waste

In 1983 FCh completed a study for the Ministry of Agriculture on availability and possible utilization of residues derived from crops, livestock, and forestry products. This study was carried out through cooperation of FCh with CORFO's Instituto Tecnológico (INTEC), a sister agency located a few blocks down the hill from FCh. An interdisciplinary group, formed by 18 professionals from both institutions, participated. The findings provided a basis for formulation of a series of high technology projects, aimed at constructive use of resources that are presently being wasted.

Forestry Waste

As part of the forestry sector studies, several waste utilization opportunities were identified and technology needs defined. In mid-1984 a mission, composed of consultants associated with the Canadian International Development Agency (CIDA), visited Chile for the purpose of ascertaining the feasibility of using forest residues as fuel in electric power co-generation. Like most FCh jobs, the mission had their terms of reference carefully spelled out.

The mission was headed by Sandy Constable, Chairman of Canadian Resource Limited (CRL), an independent consulting firm specializing in energy and natural resources. Other members of the delegation were: Peter Midgley, VP-CRL; Jens Henriksen, Bachrich Consulting; Stephen G. Gardiner, a private consultant; Carl Bojanowski, Industrial Mill Installations; and Alfonso Casasempere, a Chilean forestry engineer working with the Ministry of Industry in British Columbia (Canada).

The areas of Coelemu, Quirihue, and Chillán (Region 8) were provisionally selected as promising sites for installations. During the latter part of 1984, specialists working with FCh's Forestry Department carried out pre-feasibility studies related to three power co-generation projects.

Production of electricity for industrial purposes by means of steam operated power plants, fired by waste forest products, proved most feasible in areas producing an abundant and constant supply of such residual products as sawdust, shavings, bark, small wood, and waste from forest thinning operations. This way there were to be no interference with development of the primary resource. Any excess power would be used to supply nearby communities.

Black liquor or tar oil, the so far unusable waste product in pulp-making, is also being considered as potential raw material for a silvichemical enterprises.

Fruit

The rapid increase of exports is leaving behind ever increasing piles of fruit that neither meets export grades nor can be profitably absorbed in domestic markets. Assuming 10-15 percent rejects, this generates about 200,000 tons annually of grapes, stone fruit, apples, and pears unfit to be sold in fresh form. To do something useful with that pile of raw material is precisely the type of challenge FCh welcomes.

In 1984 FCh initiated preliminary studies aimed at identifying potential uses for fruit waste: (a) raisins made out of shattered seedless grapes; (b) apple pulp puree turned into frozen, dehydrated, juice concentrate; (c) frozen peach and nectarine sections; and (d) clear plum juice, marmalades, and canned sections and prunes.

Livestock

The operations of PROCARNE as well as other slaughter houses call for processing technology for bones, blood, guts, and skins. Likewise uses for whey, the liquid left over from cheese making, are being looked into. Waste from fish processing, wineries, corn tops, straw, and aquatic plants are also on the FCh technology search list.

MANAGEMENT OF SUBSIDIARY ENTERPRISES

By the end of 1984 the FCh's investment in subsidiary companies had grown to close to $28 million. The FCh subsidiaries were managed by their own executives. Yet it became evident that some overall management guidance would be needed so as to assure than each enterprise not only demonstrated the viability of the selected technologies, but also served as examples of effective overall enterprise management. Excellence in both areas would presumably attract potential investors and partners.

Consequently, in 1985 FCh set up a unit for management assistance to its subsidiary enterprises. The initial activities of this unit focused on help with financial management and control.

Technical Assistance and Cooperation

Technical assistance and cooperation is a crucial technology transfer activity. FCh provides aid to existing enterprises in need of a catalytic agent capable of helping to identify, select, adapts and apply more effective production and marketing procedures. Such software inputs do not call for FCh to put up any risk capital; FCh tries to sell its consulting services at cost.

FCh has increasingly been called on for tasks related to production, processing, and marketing of fresh fruits and vegetables; dairy, beef, and fish; and forestry related products and services.

Chile has soils and climates suitable for production of virtually dozens of fruits and vegetables, which are in growing demand by consumers in North America as well as in highly industrialized European countries and Japan.

Being located in the Southern Hemisphere enables Chilean farmers grow and harvest in a pattern counter-seasonal to producers in the Northern Hemisphere. This is a crucial comparative advantage, which Chile shares with producers in Australia, New Zealand, and South Africa as well as some of its Latin American sister republics.

In selecting crops as warranting technical assistance inputs for improvement of production and marketing, FCh therefore followed a simple rule: select products that can be advantageously sold in overseas markets at times of the year when local production in consuming countries is at seasonal lows. That criterion has been used in developing asparagus as well as in improving table grapes, apples, pears, stone fruits, berries, and the list goes on. The same philosophy underlies the enormously successful Quality Certification Program (QCP), reviewed later in this Chapter.

ASPARAGUS

Asparagus was introduced to Chile by Spanish conquerors. Until the late 1970s its cultivation was relegated to kitchen gardens and to a few commercial growers, producing asparagus for the local market. Asparagus was a specialty, highly appreciated by the Chilean "gourmet set."

Early in its activities, FCh saw promising export markets for the crop and decided to look into marketing opportunities for green asparagus in West European and U.S. markets. In 1977 FCh brought in two American specialists to help start pilot cultivation.

FCh assured potential growers that outlets for their crop would be available, and gave them a Spanish copy of its market research study.

Since the colonial days the "Mary Washington" variety was almost exclusively grown in Chile. It is suitable for white asparagus. Export markets require green asparagus. FCh brought in the UC72 double purpose variety: white spears have to be cut while underground; once the spears grow out of the ground the sun turns them green.

FCh provided technical assistance to 40 small asparagus growers, who obtained production credit from the Agricultural Development Institute (Instituto de Desarrollo Agropecuario, INDAP). FCh provided initial quality plants, grown in its own nursery. Delivery of seedlings was accompanied with a technology package including advice on soil preparation, planting dates, and an annual program of work, all tailored to the possibilities of individual growers. FCh professionals did field checks on management practices, visiting growers about eight times a year.

Apart from the technical assistance to pilot growers, an agreement was signed with INDAP to introduce the crop to small farmers.

In short, FCh strategy was to play both sides. The recommended technology for white asparagus is to put seedlings a little deeper, so that plants can be hand cut underground. For green asparagus, seedlings are given a more shallow planting and mechanically harvested.

FCh studies indicated that fields with 20,000 plants/ha, mechanically harvested, could gross $3000/ha and generate an internal rate of return (IRR) of 22 percent for the green as well as white type of asparagus. The crop appeared economically attractive. As a result, acreage went up steadily, exceeding 1200 hectares in 1983. Production is concentrated in regions 5, 6, 7, and 8.

FCh technical assistance included development of post-harvest technology, quality control, and export promotion. Exports increased from practically nothing in 1979 to 200 tons in 1982, and were expected to quadruple to 800 tons in 1983. It did not turn out quite that way. Of the 125 tons exported, one-fourth was rejected in U.S. ports because non compliance with phyto-sanitary standards (i.e., insect infestation). However that problem could be readily resolved with suitable preventive measures: incorporation of insecticides in the soil and weed controls.

By late 1984 asparagus was back on track. During September December Chile air shipped some 170,000 boxes (about 1000 tons equivalent), with over four-fifths destined for the United States. By 1986 output approached 4500 tons.

Hernan Monardes, the technician responsible for asparagus work at FCh, developed modified atmosphere containers to help transport fresh asparagus by sea at a fraction of the air freight costs that•are currently being incurred during peak season.

Usually only about half the harvested crop meets standards for exports in fresh form. FCh is therefore exploring ways to process the other half of the crop by means of suitable adaptation of existing canning, freezing, and drying technologies. The outlook is promising for increasing total revenue from this integral utilization of the asparagus crop.

In short, asparagus seems a success story.

DAIRY TECHNOLOGY

The initial FCh diagnostic study of the dairy sub sector showed that: (a) in central and southern Chile there are ideal pasture conditions for production of milk; (b) the per capita consumption was only about 83 liters (half in fluid milk, powder, and butter; the other half in cheeses); (c) some 80 percent of dairy cattle are located in southern Chile, while the bulk of consumers live in the Santiago Valparaiso metropolitan areas; (d) milk production and quality fluctuates with season and distance from market; (e) dairy products from distant farming areas tend to be contaminated, resulting in short shelf lives and highly variable quality; (f) by-products such as whey (suero) were being dumped into lakes and streams, causing pollution and wasting a nutritionally valuable product; and (g) unfavorable fluid milk prices made more and more of the small 100-odd estate (fundos) dairies switch to artisanal cheese (chanco), produced under unsanitary conditions, resulting in variable quality and low prices.

In order to persuade companies that improved technologies could really help them individually improve their "bottom line," FCh had first to get a foot in the door. Thus in 1976 a functional plan was elaborated with three major thrusts: (a) technical assistance aimed at improving product quality, (b) new product development, and (c) use of by-products.

Quality Improvement

The project had to be medium- to long-term, with effort to be sustained over at least 3 years. An introductory seminar on food industry sanitation opened the way to intense studies of the Chilean dairy industry. Consulting arrangements were made with several cooperatives and private cheese producers. FCh provided basic services for chemical analyses of pesticides and antibiotics as well as microbiological testing and sensory analyses.

FCh came to Chile when the dairy industry was in the process of major structural changes. Shake-outs and mergers were reducing the number of larger multi-product dairy plants from 35 in the late 1970s to about 12 in mid-1983. This left several regions without access to milk plants or with a monopoly of one plant. Due to the economic crisis and shortage of foreign exchange, importation of specialized dairy products, including cheese, ceased. This opened up possibilities for import-substitution by small local plants.

FCh did not stop at improving the dairy products alone. Taking an integrated view of the technology process FCh wanted to make sure that consumers could be suitably advised by attractive packaging and point-of-sale (POS) displays.

Ultra-High-Temperature Processing

FCh furthered adaptation of ultra-high-temperatures (UHT) milk processing. UHT makes it possible to decentralize large pasteurization plants away from the Santiago metropolitan area and other urban centers. This is because UHT milk has a shelf-life of up to 6 months without refrigeration; it can therefore be transported to remote areas in the south of Chile. There wholesome milk can be made available to consumers, who previously had to resort to reconstituting powdered milk.

Growth of UHT use in bottled milk has been phenomenal, increasing more than eight-fold between 1977 and 1983, as shown in Table 3.8 (in millions of liters):

Table 3.8 Diffusion of Ultra-High-Temperature Milk Processing, Chile, 1977 and 1983

Factor	1977	1983
Pasteurized	119	61
UHT	6	52
Total	125	113
UHT as % of Total	4%	46%

Product Development and Adaptation

Introduction of new products required not only food technology transfer but also use of modern marketing techniques to identify opportunities for successful placement in the marketplace.

In mid-1985 FCh invited S. Gantchi, a Swiss merchandising expert, to demonstrate in-store methods for display of dairy and meat products in refrigerated cabinets of self-service stores.

Ementhaler cheese was thus launched with Swiss know-how. Parmesan cheese was introduced with Italian know-how, identified during a visit of a FCh staffer to the University of Wisconsin in Madison.

Choice of an appropriate project manager was critical to marketing project ideas to local enterprises. Jorge Hermanns, a Chilean veterinarian with production experience in the then-largest dairy cooperative, was chosen. Work was initiated by: (a) publicizing the practical value of FCh; (b) identifying technologies that needed to be transferred; and (c) defining the type of assistance that could be provided by the Milk Technology Center (Centro Tecnológico de la Leche, CTL) at the Universidad Austral.

In addition to quality control, FCh geared up to provide three specific services: (a) solve mechanical and contamination problems in UHT equipment; (b) evaluate packaging and equipment; and (c) introduce Swiss-type yogurt as the type that fit Chilean consumer preferences. As a result of the product improvement the

Unión Lechera de Aconcagua in Viña del Mar, a large dairy plant, was able to increase sales from 180,000 units/month to 560,000 units/month in 4 months without doing any advertising.

Under auspices of the dairy project, other achievements have been realized: (a) nine new products have been adapted to specific conditions of the Chilean market (spread cheese, white cheese, whipping cream, camenbert cheese, standard yogurt, a sundae-style yogurt, puddings, flan, and a whey beverage); (b) cottage cheese has been adapted from the United States and is now being sold successfully in increasing quantities; (c) other new items include custards, puddings, and a Swiss-style yogurt, which uses many varieties of excellent Chilean fruits; and (d) amply available goat milk was channeled into improved varieties of quality goat cheeses, an activity that generated enough consumer demand to justify setting up a manufacturing corporation with seed investment contributions from FCh and local agricultural cooperatives (see section on CAPRILAC).

In all these innovations FCh's well-equipped pilot plant played a key role in technology transfer, as well as source of production sufficient for consumer and market testing (ultrafiltration unit, UHT, Tetra-Pak containers, cheese vat, pasteurizer, centrifuge, etc.).

A great deal of progress has been made by FCh in aiding dairy producers to compete more aggressively by helping them effectively absorb and adapt suitable technologies. Yet by the mid-1980s the Chilean dairy industry was not over the hump. Far from it. There were still such challenges ahead as competition from inexpensive imported powdered milk; the beef industry, which buys up dairy stock for slaughter due to more favorable meat prices; and fruit and vegetable growers competing for dairy land, particularly in the surroundings of Santiago.

Initially FCh worked with large companies simply because they needed access to foreign sources of technology. Several of the contacts that which FCh helped to catalyze resulted in joint ventures: Borden, with Lechera del Sur and Beatrice Foods with Los Alamos. These arrangements institutionalized transfer of technology within the framework of joint ventures, FCh continues to provide laboratory services to large companies, who do not have specialized equipment needed for such analyses as inhibitors, insecticide contamination in milk products, checks of their own laboratory, or establishment of performance and quality guarantees for raw material.

Improvement of Small Cheese Plants

In 1984 Chileans consumed 33,148 tons of cheese; 60 percent was made by small artisanal dairies.

Table 3.9 Cheese Supply, Chile, 1984

Origin	1000 tons	% of total
1. Industrial dairies	12.4	38%
2. Artesanal dairies on agri- cultural estates (fundos)	20.0	60
3. Importations	0.7	2
Total	33.1	100%

Over 90 percent of the industrial cheese production came from six enterprises: La Unión, Los Alamos, Loncoleche, Lechera del Sur, Bío Bío, and Valdivia. These plants were reasonably well equipped and had technically competent staff. In contrast, some 20,000 tons, or 60 percent of the 33,000 tons of cheese consumed annually in Chile, came from the 100 or so small dairies, receiving less than 10,000 liters of milk daily. These artisanal enterprises made local cheese, called chanco, using raw milk of widely varying quality. As a result no two chanco batches ever came out alike. The differences between winter and summer chanco were particularly large. Consumer prices, however, were low.

FCh saw an opportunity for making a substantial impact by introducing relatively simple changes in production technology. The first step in quality improvement was pasteurization of milk. FCh helped develop a starter that would produce a pasteurized cheese, similar to traditional chanco. FCh assisted small plants in the Santiago metropolitan area to modernize equipment with introduction of stainless steel tankers, washable floors and walls, improvements in maturation chambers, ventilation, and control of rodents and insects. The combination of small changes in technology made it possible to lower use of raw materials in cheese making from 12 liters of milk per 1 kg of cheese to only 9 liters of milk per kg of chanco. The resulting whey, formerly not used, was being utilized.

The trade was soon willing to pay a substantial premium for the FCh type of chanco sold at the weekly auctions held at the Tattersal Product Fair in Santiago, which is the most important national wholesale market for grain, hay, charcoal, potatoes, legumes, nuts, almonds, and cheese.

In 1983 FCh expanded its technical assistance to producers of such exotic cheeses as camenbert, eidam, gouda, gruyere, rochefort, and parmesan. Pilot work on Camenbert indicated that the local cheese could be placed in retail outlets at a price of 40 percent below imported cheeses.

The battle was not quite won, however. By the time chanco reached retail stores the price has doubled from the Tattersal levels, but the point-of-sale display tended to be so inadequate—dark and dirty display cabinets, filled with haphazard assortments of irregular cuts of dried out and unattractively packaged cheese cuts—that the improved chanco would not command the price premium it deserved. FCh felt that its job was not done unless retailing practices were improved as well. Consequently, in May 1985 S. Gantschi, from the Swiss Milk Industry Advertising Center, was brought to Chile for the purposes of transferring cheese marketing technologies.

The combination of better yields and prices rapidly paid for the investment in equipment. Sample budgets from small client dairies indicated internal rates of return (IRR) ranging from 22 up to 66 percent.

DON PEZ

Paradoxically, in face of all the abundance, domestic consumption of fish was low and declining from 6.3 kg/capita in the early 1970s to 4.6 kg/capita in 1982. This compares to about 40 kg/capita in Japan. One reason is that fresh fish for domestic use has been marketed by archaic methods. The main constraint appeared to be the distribution system, which failed to provide ready access to reasonably priced, fresh quality fish for consumers living away from coastal areas. In 1983 the Subsecretariat of Fisheries entrusted FCh with development of a national distribution system for fresh fish, to be based on a pilot operation in Coronel (Region 8). The project is called Don Pez (Mr. Fish).

Don Pez focused on maintaining an unbroken cold chain from catch to market by special fresh fish handling, preparation, and packing methods. Use of the FCh designed technology permits a sanitary handling of fresh fish and extends shelf-life to over a week. This integrated network of catch, processing, and marketing opened reliable market outlets for artisan fishermen as well as an improved supply of quality protein in the Chilean diet. This is especially important during summer months, when consumption of fish in Santiago is at seasonal lows.

Similarly to the case of PROCARNE, the Don Pez program required changes at each stage of the vertically integrated system so as to protect the quality of the attractively packaged product. FCh specialists, with occasional help from overseas technicians, conducted intensive training of artisanal fishermen, packing house workers, and personnel in supermarket and other retail outlets.

By 1985 it became evident that the program was a success: 40 fishermen participated in development programs aimed to expand sales of fresh fish in supermarkets of Metropolitan Santiago. This involved rigid quality control linked to rapid, sanitary, refrigerated transport and display. By 1986 FCh started to expand distribution to institutional clients.

Judging by the sales upsurge, to Don Pez brand of fish is helping to overcome the poor quality reputation the product traditionally had in Santiago markets, about 90 miles inland from the coast where the fish are caught. Don Pez might well prove to be one of the most cost effective transfers of marketing-related technologies FCh has so far undertaken.

Minced Fish

In 1978 FCh initiated a program aimed at using fish directly as food ingredients, rather than manufacture fishmeal for export. Miguel Yoma, of Pesquera del Sur, with help from a Japanese scientist, demonstrated the value of using minced fish from fish parts and species not usually used for food. The raw material served as ingredients in making high grade sausage, puddings, cakes, and croquetes, which were successfully consumer tested. However, competition from sausage made with such low-cost material as chicken, plus the depressed state of the Chilean economy, made it impossible to raise the capital needed for implementing the project. It has, therefore, been shelved and is ready to go whenever adequate funds can be mobilized.

LUPINE

Since World War II, legumes, other than soybeans, have generally been subjected to benign neglect in production as well as research. That is a pity because—in contrast to traditional food crops, which require high inputs of energy, technology, fertilizers, and pesticides—some legumes can be grown under marginal conditions in rotation with other crops.

Lupines are a case in point. They: (a) are frost resistant; (b) at 3-6 tons/ha, out-yield most other legumes; (c) serve as a soil conserving agent; (d) adapt to a wide variety of soils and climates ranging all the way from semi-arid regions of Australia to southern Chile, with 1800 mm of annual rainfall; and (e) grow at sea level in Italy as well as at 3800 m altitudes around the Titicaca Lake in Bolivia.

In 1979 FCh signed an agreement with the Experimental Station of the University of Chile in Carrillanca to review results of ongoing research on lupine; define additional areas of information gathering; and outline action recommendations. The Lupine Information Secretariat (LIS) was set up to organize annual meetings for representatives of universities, technical institutes, experimental stations, oil milling, and biscuit and food processing industrialists. By 1981 FCh was able to withdraw and let the lupine work run its course, with LIS providing work coordination.

CENTER FOR TECHNICAL ASSISTANCE FOR THE CANNING INDUSTRY (CAT)

Studies carried out by the Food and Agriculture Organization of the United Nations (FAO) in Chile during the 1960s identified major problems that hindered orderly development of the canning industry: (a) poor quality equipment, lack of maintenance, and high cost of spare parts; (b) poor plant sanitation; (c) poor and varied raw material due to lack of quality specifications and control; and (d) lack of knowledge about sources of information on process technologies, including suitable manufacturing equipment.

The going was rough: (a) CAT managers underestimated the long-term difficulties associated with changing behavior patterns among members of the industry; (b) FCh staff members assigned to the CAT project had previous experience in US trade associations and prepared work plans on basis of successful U.S. precedents; (c) project teams did not succeed in actively involving a large enough portion of the industry; (d) CAT was unable to pass on to industry a significant share of its costs for development and maintenance; (e) canning companies tended to be secretive and reluctant to pay for programs that they would have to share with competitors; (f) quick returns on investment were expected, while CAT was offering long-term benefits derived from collective action, without dramatic short-term benefits to individual firms; (g) canners perceived the opportunity for exporting "too remote" to justify active involvement with CAT; (h) few scientists in local academic institutions were excited by promise of unglamorous work related to improving manufacturing practices; (i) the economic boom of the late 1970s did not place pressure on companies to dramatically improve or adapt their technologies.

In short, few canning companies were committed to significant technical change. CAT was therefore discontinued in 1982.

AGRO-ECONOMIC INFORMATION SYSTEM (SIAE)

Until the early 1980s Chilean farmers did not have access to elementary data on yields, costs, and prices to facilitate decision making about what to grow, when, and how much.

In the early 1980s the Office of Agricultural Planning (Oficina de Planificación Agrícola, ODEPA) initiated work on preparation of an Agricultural Sector Model. The University of Chile and Catholic University joined with FCh for that purpose. Dr. Robert Echeverría, a Chilean working at the World Bank in Washington, D.C., was seconded to the project to help develop the methodology.

FCh then entered into a series of agreements with regional sector organizations aimed at providing location-specific information that would help alleviate this data drought. In 1984, FCh joined with the Institute of Agricultural Economics at the

Austral University to initiate a pilot Agro-Economic Information System (SIAE) for Region 10.

The specific purposes of SIAE were to obtain and analyze the best available information related to the following subjects: (a) definition of agro-economically homogeneous areas in terms of soils, climate, water, market conditions, and other economic factors; (b) production costs of farmers in each homogeneous area for major products of the area; (c) yields; and (d) expected prices for major products of the Region 10.

The resulting analysis made it possible to make estimates about profitability of different crops in the year to come.

The pilot data appeared twice a year: in February to guide decisions about crop planting and livestock breeding to be made in the winter; and in July, to guide production decisions for spring.

FCh obtained cooperation from major regional agricultural sector organizations: (a) Region 10 Planning Office, (b) Agricultural Society of Valdivia (SAVAL), (c) Agricultural, Dairy, and Consumer Cooperative of Frutillar (CAFRA), (d) Agricultural and Livestock Society of Osorno (SAGO), and (e) agrobusiness enterprises (Auction Market in Osorno, Grob Mills, Angelini-Grace, Indus).

The cooperating agencies helped FCh to divide Region 10 into five homogeneous areas: (a) coastal mountains, (b) coastal foothills, (c) central valley, (d) the Andean foothills, and (e) the Andes.

The following 4 product type were included in the pilot SIAE run: crops (wheat, oats and mustard); milk; livestock (beef, fall calves, breeders 6-18 months, fattening heifers 18-30 months, and slaughter dairy cows); and pastures (natural, improved permanent silage and hay).

In the introduction to the first SIAE report, FCh pointed out that the Informativo Agro-Económico is "only applicable to Region 10 in terms of yields, crop management, used inputs and costs... the data being only a guide to the farmer, and do not mean to be a forecast of what will occur in the future [because] many uncontrollable factors affect yields, costs, and markets, which would bring about results differing substantially from FCh indications of profitability...."

The simplicity, relevance, and timeliness of SIAE hit a responsive cord among target audiences of farmers in Region 10 and beyond. By late 1985 SIAE reports were prepared for three more regions: 5, 6, and 7.

The periodic issue of SIAE reports gave FCh an opportunity to continual remind agriculture-related publics of its presence and purpose. Furthermore, SIAE reports also brought in inquiries about additional work for individual clients, thus becoming a marketing for FCh services.

KIWIFRUIT

In the early 1980s kiwifruit started to be grown in Chile. By 1985 Chilean farmers had 800 ha (2000 acres) of kiwifruit plantations, producing 16,000 tons. The Chilean fruit matures 2-4 weeks before that from New Zealand, the major Southern Hemisphere producer.

Kiwifruit is among the most profitable crops in Chilean agriculture. About 70 percent of the crop is fit for export in fresh form. Producers stand to improve the profitability of the crop even further, should they find a way to utilize the 30 percent of kiwifruit that does not meet export standards.

FCh took on the task of adapting technology for processing reject fruit into pulp, wine, juices, and jams. In June 1985, as an initial step, FCh organized a short course on production, management, and processing of kiwifruit.

QUALITY CERTIFICATION PROGRAM (QCP)

In Third World countries farmers and artisans offer for sale products that vary in size, quality, and content. Before buying, consumers therefore inspect goods one by one, a very time consuming transaction. In contrast, modern mass production and marketing is based on the ability of manufacturers to deliver to buyers batches of merchandise of known and uniform quality. A vigorously enforced system of mutually agreed on grades and standards makes it possible to quickly make sales transactions without inspecting individual batches. Much of the trading is done on a "sight unseen" basis.

In the late 1970s FCh realized that the inability of Chilean shippers to maintain a high quality of merchandise was among the major constraints slowing growth of exports of fresh produce. The evidence was readily at hand.

In the European and U.S. markets Chilean fruit was bringing prices well below the carefully-graded fruit from Zealand, Australia, and South African competitors. Under those circumstances FCh recognized a unique opportunity for transfer to Chile of established standardization, grading, and quality control technologies for fruits. The resulting Quality Control Program (QCP) has since become one of the "flagship" activities of FCh. Its FCh red-and-blue quality certification seal has become well known and accepted by produce importers overseas. This cost-effective program is rapidly expanding to table grapes, vegetables, dairy products, seafoods, wine, forestry products, and so on. There seems no end in sight.

Fruits

The QCP began modestly enough in 1980. FCh started working with farmers to assure their crops would be produced correctly. FCh's technical assistance thus provided fruit growers and cooperatives with up-to-date management information. Field assistance was complemented with courses, seminars, and literature. Initially some 20 apple growers, with about 1000 ha of orchards, signed up. Typically, two field visits by FCh personnel were made each year.

By 1981 FCh was ready to move into a post-harvest program with a few shippers, including the Chilean subsidiary of the Standard Fruit Corporation (SFC). During 1983 FCh experts were sent to main European ports to observe the conditions in which Chilean fruit reached those markets, and to watch handling right to the moment of purchase by final consumers. The experience gained enabled FCh to structure an integrated Quality Certification Program (QCP).

During 1982-83 six companies signed up for QCP. The initial program consisted of two major components: (a) quality inspections at packing and shipping sites in Chile, and (b) quality certification to satisfy standards at such European terminal ports as Hamburg and Rotterdam, where an FCh inspector observed the condition of fruit upon arrival to assure that undue damage did not occur in transport.

In 1983 QCP was bolstered by the Chilean government's passing an Export Quality Certification Law.

The number of boxes of fresh fruit inspected by FCh has increased from 4.5 million during the 1981/82 season to more than 12.0 million boxes during 1984/85, representing one-fourth of the 10 million cases of all fresh fruit exported from Chile:

Table 3.10   Fresh Fruit Inspected in Chile, 1981-85/in millions of cases

Year	Total Shipments	Quality Certified by		FCh as % of Total
		FCh	Others	
1981/82	28.0	4.5	–	14%
1982/83	34.0	7.4	8.0	22
1983/84	41.0	10.5	12.0	25
1984/85	50.0	12.0	N.A.	24

The charge for certification was an equivalent of about U.S.$0.03 per box. This basically covers FCh's direct costs. Benefits derived from the greater reliability of Chilean fruits and vegetables accrued not only to producers and exporters, but the whole country.

An increasingly important complementary activity is FCh's control of pesticide residues on export fruits. It aims to meet requirements of importing countries. With prohibition of cancer-related diethylbromide (DEB) fumigation, the quality certification program is moving into the search for safe fruit treatment technologies, such as irradiation and hot water dip.

During the harvesting and shipping season, some 45 specialists, who have been trained specifically for the job, work in regions 3 to 7. They are in packing stations and at ports, inspecting fruit in accordance with quality standards laid down by individual companies or importing countries, trying to solve problems likely to arise following inspection. Quality affidavits, which accompany individual shipments, certify that exported fruit meets with specific requirements or standards agreed between exporter and buyer. Consequently, advance payment can be made against the guarantee that is provided by the certificate.

In 1985, in an attempt to further improve the service, FCh surveyed a sample of U.S. buyers to find out whether inspection procedures satisfied customer requirements. The majority had no complaints; they wanted FCh to increase the scope of the work. As a result, during 1987/88 FCh plans to add routine checks for traces of insecticides at the level of fruit plantation.

By 1985 the quality program has taken on such importance in the fruit export trade that the Ministry of Agriculture established a National Committee to guide the work. In addition to FCh the Committee includes representatives from and/or actively cooperates with producers, universities, the Institute for Agricultural Research (INIA), the Export Promotion Council (PRO-CHILE), the Agricultural Planning Office (ODEPA), the Crop and Livestock Service (SAG), and the National Agricultural Association (SNA).

FCh is vigorously advertising its successful QCP. Attention-getting "Buy Quality/Buy Profit" ads for the FCh Quality Certification are being placed in every issue of the International Fruit World, the leading trade publication. The text then says how: "We inspected and certified for export over 25 million boxes of Chilean fresh fruit during the past three years. Growers and exporters are satisfied, but the happiest of all are the importers because they received better prices in the market place. When you buy Chilean fruit, you get excellent produce. When you ask for Fundación Chile quality certified Chilean fruit, you are guaranteed the absolute best. Request our Quality Control Certificate... it means PROFITS!"

Nothing succeeds as well as success. By 1984 FCh expanded its quality control program to fresh vegetables, fish, shellfish, wine, and forestry products. By 1985 it started to work on wood and processed fruits and vegetables as well.

Seafood

During 1983—similarly to the adverse experience with fresh fruit exports—some $200,000 worth of canned Chilean shellfish were returned from the United States; they did not meet sanitary standards of the Food and Drug Administration (FDA). If similar difficulties were to be avoided in the future, QCP evidently needed to be extended to seafood. As a first awareness-raising step, FCh, in cooperation with the Chemistry Department at the University of Santiago, organized a seminar for sanitary supervisors of fish processing in Chilean plants. The November 1984 event was supervised by Dening M. Dynan, an FDA food technologist.

It soon became evident that leading processors already operated under quality control systems. The 15 fish meal and fish oil plants affiliated with the Fishery Corporation (CORPESCA)—which in 1984 accounted for 80 percent of national production, and $250 million in exports—had their own quality control departments. Their reports were being certified by CORPESCA, according to standards set by Lloyds, Veritas, and other major insurance companies.

Likewise, in canned fish Sociedad Pesquera Coloso, one of the principal exporters, maintained quality control from catch to organoleptic and bacteriological analysis of finished products. Coloso focused on freshness, size, cleaning, gutting efficiency, portioning, weight of raw canned product, preparation of cans, time and temperature during sterilization, temperature of cooker, exit temperature of cans, and can-seal analysis.

Pesquera Yadran, a leader in frozen and canned shellfish, (sea urchins, abalone, mussels, clams, crabs, and razor clams) exports mainly to Japan, the United States, Taiwan, Hong Kong, and Western Europe. Quality control applies to water supply, hygienic and sanitary services, waste water disposal, as well as to finished products. Finished products are being submitted to quantitative research of microorganisms, weight control, quality of packaging materials, and specifications contracted for by buyers.

All that FCh needed to do was to select the best of existing local practice, fill in missing elements, and consolidate it into an enforceable system. In mid-1984 a direct and explicit recognition of the value of FCh quality certification of fish products for export came from a most unexpected source. The Insurance Company of North America (INA) started to issue comprehensive export coverage at lower than standard rates to merchandise inspected at processing plants by FCh and carrying FCh certificates for individual shipments.

A 1984 study of the South African Market for Chilean fishery products indicated that: (a) canned fish was being primarily consumed by the 20 million low-income people who make up almost four-fifths of the total of 26 million of inhabitants; and (b) South African fish catch declined catastrophically from about

2 million tons in the late 1960s to about 0.6 million tons, a
level at which it stagnated since the mid-1970s.[1] Evidently
there was a large potential market for Chilean exports to help
substitute for declining South African catches.

In 1985 FCh negotiated a very specific certification
agreement with the South African Bureau of Standards (SABS),
which is responsible for enforcement of sanitary, quality, and
appearance standards established for imported products. Without
SABS seal of approval products may not be marketed for human
consumption in South Africa.

The SABS/FCh certification is simply a guarantee that,
before leaving Chilean shores, the products meet pre-established
standards of uniformity, size, color, appearance, texture,
filling of containers, ingredients, labels, weight, modes of
processing, and origin of raw material.

Wine

Climatic conditions in central Chile are favorable to
development of vineyards. As far back as in 1548 the first vines
were planted by Jesuit priests. During the 18th century, Chilean
travelers returning from Europe brought back noble varieties of
grapevine stock from France.

In 1985 production was about 0.5 billion liters of wine from
some 100,000 ha of vineyards. There were virtually hundreds of
wine brands competing in the domestic and export markets.

The best known brands come from 14 vineyards (Concha y Toro,
Cousiño-Macul, San Pedro, Santa Rita, Santa Carolina, Ochagavía,
Undurraga, Tarapacá, Linderos, Cánepa, Santa Helena, Carta Vieja,
Carmen, and Miguel Torres). With a combined 1985 storage
capacity of about 1.2 billion liters, these wineries can store 2
years' production.

Fine wines accounted for about 155 million liters; some 85
percent of sales were concentrated among the big four wineries:

Table 3.11 Production of Major Chilean Wineries, 1985

Winery	millions of liter	% of total
1. Concha y Toro	44	28%
2. San Pedro	33	21
3. Santa Rica	33	21
4. Santa Carolina	22	15
5. Others (Santa Elena, Tarapacá, Cánepa, Cousiño-Macul and Ochagavía)	23	15
Total	155	100%

In the early 1980s Chile exported about 15-20 million liters annually, or 3-4 percent of total production.

The bulk of Chilean exports were of dry, woody-tasting red wines of the Cabernet-Sauvignon type. FCh felt that exports could be increased substantially if Chilean vintners would improve the quality of their product. Consequently, in 1983 FCh brought to Chile Dr. John Hoffman, a prominent U.S. wine expert, to assess the state-of-the-art in winemaking. Dr. Hoffman concluded that Chilean vineyards were well managed but that, in contrast, inadequate grape processing technology was being used by small vintners. Production technology in white wine was deemed to be about 30 years behind the technology currently used in California. Dr. Hoffman then devised a Technology Transfer Program for Wine (Programa de Transferencia de Tecnología de Vinos, PTTV). PTTV is an action plan for quality certification of Chilean wines, aimed at expanding penetration of international markets by Chilean quality wines.

The quality certification program was based on precedent from several Chilean vineyards, who have voluntarily maintained high export standards, including Concha y Toro, San Pedro, and Santa Rita.

The PTTV quality certification program went hand-in-hand with FCh's technical assistance to small vintners, which aims to enable them gradually to enter export markets. The emphasis was toward fruity, white and light quality wines, rather than on competing with mass-produced table wines. The FCh certification label on wine bottles includes name of variety, harvest year, and origin of grapes. By early 1985 the FCh seal was put on bottles from nine wineries: Astaburuaga, Monte Alto, Del Pedregal, Portal del Alto, San Ignacio, Miguel Torres, Santa Mónica, La Fortuna, and Don Francisco, all from Region 8. Specifically, the wine QCP is to help vintners adapt their products to market requirements, and boost exports to about 6-8 percent of national output by the early 1990s.

Pine Boards

By 1984 FCh—together with Biobio University in Region 8—prepared standards for grading radiata pine boards.

Prefabricated Housing

Early in 1985 FCh established a seal of quality construction of energy efficient wooden housing (sistema de construcción energitérmico).

In contrast with all the rest of the quality seals that are being issued by FCh, this one is sponsored by an Inter-Institutional Consortium of the University of Chile and

FCh. The promotional booklet announcing the new program highlights "the ample experience of these institutions with transfer of technology and materials testing, its international contacts with institutions that are worldwide leaders in subjects such as quality control, wood treatment, efficient use of energy and structural designs." This consortium offers: specialized technical assistance to producers of components and construction enterprises, as well as certification of quality finished housing.

In summary, FCh's Quality Control Program (QCP) is probably its most cost-effective project. Its success is not due to bringing in new technology. Standards of the USFDA and other agencies in importing countries were known in Chile and partially reflected in the work of the Instituto Nacional de Normalización (INN). Yet would-be exporters, acting on their own, found that specifications requirements for individual products were hard to come by. The sources of information were widely dispersed among such entities as Pro-Chile, Agricultural and Fisheries Extension Services, Exporter Associations, as well as commercial attachées at embassies of importing countries. The success of QCP lies in FCh's ability to consolidate the knowledge in one place and make the system operational as well as profitable for exporters.

ELECTRONICS AND TELECOMMUNICATIONS

In addition to all the above described food and agriculture related technologies, FCh also entered microprocessing and rural telephony, both activity areas in which ITT had much to offer.

Computerized Process Controls

The priority need of Chilean enterprises is modernization, full or partial, of plants so as to adapt them to new requirements of production, cost, safety, and ease of operation.

Microprocessors are key components of single board computers, some of the most reliable and least costly hardware. The policy of FCh is to add software value to the system by creative engineering aimed at developing automatic control systems for complex industrial processes using microprocessors and industrial microcomputer technology. In 1979, responding to the requirement for software, FCh created a microprocessor unit, charging it with 5 specific tasks: (a) identify needs for industrial control in major national enterprises—Refinería de Petróleo de Concón; Compañía del Cobre (CODELCO), Division in Chuquicamata, El Salvador and El Teniente; Minera Disputada de Las Condes; Empresa Nacional de Petróleo (ENAP) in Magallanes; Standard Electric Company; (b) design integrated systems based on rigorous selection of standard hardware and well known microprocessors; (c) develop specific programs with high-level, tailor-made language; (d) transfer technology to technicians of

FCh client enterprises; and (e) provide integrated support for locally installed systems.

The microcomputer applications initially put into practice fell into four major groupings: (a) systems for data acquisition and tele-measurement; (b) remote transmission of orders and remote control; (c) automatization and automatic controls; and (d) telecommunications systems.

A typical example of FCh work was done under contract with the petroleum refinery in Concón. The first job was to develop a radio based remote controls for pumps and wells, instructing replenishment of water losses in the cooling system, which requires 2000 liters/second. Due to the crucial importance of the task, the equipment was developed with substantial backstop hardware and permanent monitoring so as to maintain uninterrupted operations.

On basis of that experience FCh extended control and automatization to the entire cooling "loop." This called for design of a remote control for all the cooling towers that would continuously inform operators, by visual monitors, about the status of the entire system. The operator can stop or start ventilators, empty tanks, and perform other tasks supporting the cooling systems. As a natural complement of those controls an alternative system of spare pumps, which feed the replenishment water, was installed at the refinery. Likewise an emergency control system was designed, which permitted automatic starting of pumps and communication channels for water needed to successfully resolve a potential disaster.

Other examples of FCh microprocessor activities are development and implementation of systems for: (a) remote measurement and control of transport and grinding of minerals; (b) collection and analysis of oceanographic and meteorological data used for operation of oil drilling platforms; (c) automation of security measures at a petrochemical plant; (d) automation of retail sales of fuel at service stations, and so on.

In short, FCh technicians were able to specify, relatively quickly, suitably dimensioned systems at low costs and with appropriate technological capacity. These procedures were contributing substantially to cost effective improvement of national industries.

In 1986 FCh expanded its activities to include mass producing multi-purpose computer systems specifically designed for applications in automation of industrial processes. It took eight years of development and adaptation work for FCh to introduce the Integrated Control System, known as SIC-15. CORFO referred to SIC-15 as "landmark in Chile's engineering history... targeted to satisfy automation requirements of a number of industrial processes, for which imported systems often turn out to be oversized, too expensive and/or inflexible." [2]

The SIC-15 "opportunity window" was identified in 1978, shortly after FCh initiated operations. It was then and remains now clear that it would be difficult for Chile to develop computer hardware. In contrast, FCh sensed that there seemed to be opportunities for developing software packages tailored to specific industrial applications. This ran against the tendency to use computer systems primaraily in routine administration and accounting.

A small group of U.S.-trained FCh professionals set out to exploit this market segment by aiming to: (a) build into SIC-15 top quality design; (b) modular structure that would keep down costs, yet readily configurate the system so as to suit location-specific requirements; and (c) complement the basic package with local consulting on hardware and software as well as maintenance upgrading and special application support.

SIC-15 was designed to be a user friendly, highly reliable, and readily expandable system with a broad range of flexible applications ranging all the way from simple and routine data gathering to large scale industrial control and monitoring processes.

SIC-15 is built around a central communications node microcomputer, typically an Intel 8088 microprocessor, which can be interconnected with up to 16 remote devices such as IBM PC XTs or ATs, as well as 16-bit Intel 8088 microprocessors. As an integrated system, interconnected to a diversity of other types of systems, SIC-15 allows the user "to distribute eggs in more than one basket." This helps hierarchies control, distribute intelligence, and reduce risk of major mishaps—which are all serious shortcomings of single mainframe computers.

The 1986 release of SIC-15 simply synthesizes and augments 14 highly successful, custom-designed industrial control systems, such as Concón oil refinery radio-telecontrol, CODELCO's privately owned electronic mail, ENAP-Magellan oceanographic and meteorological system, autogenous grinding operation at the Disputada copper mine, and so on. FCh can demonstrate that integrated control systems really work in today's Chile—a powerful sales argument.

In short, SIC-15 is a high-technology system developed in Chile and designed for the country's industrial process control needs. Its low-cost full local technical support makes it particularly attractive to mining, forestry, petrochemical, and chemical industries. Without FCh's early and energetic involvement with transfer and adaptation of SIC, Chile simply would have advanced at a far slower pace with its application.

Rural Telephony

Rural telecommunication is another FCh activity, closely related to agriculture and within the framework of traditional ITT technology.

Most small Chilean farmers have no telephones, greatly hindering communications with the outside world on such life-and-death matters as health related emergencies.

Furthermore, lack of instant access to current market data prevents farmers from effectively negotiating sale of their products or comparative shopping for farm supplies.

An initial FCh survey, made in the late 1970s, showed that there were about 800 rural localities of up to 1000 inhabitants without telephone service; and due to low profitability of such projects, chances were slim for farmers to obtain this service in the foreseeable future.

Rules of the game in Chile do not allow for subsidies to help development of rural telecommunications. Telephone densities in certain areas, such as the extreme southern provinces of Aysén and Magallanes, are less than one subscriber per 400 km^2. This allows for at most one conventional public telephone to serve the whole area. Some nonconventional approaches were evidently needed if telecommunication coverage in rural areas was to be intensified. Consequently FCh took the initiative in joining with community members, local government authorities, and operating telephone companies for the purpose of identifying technologies capable of providing service to these localities.

In 1979 a group of 25 farmers formed a telephone cooperative in La Herrera farming locality, 20 km south of Santiago. The purpose was to operate a small network with a 40-line PAB, the first use of this technology in Chile. Total cost per subscriber—including switching, power, metering, trunking, and outside plant—came to U.S.$1700, or about the order of magnitude charged for installation of a residential phone in Santiago.

In 1982 the largest rural network in Chile was inaugurated in Llanquihue, 1000 km south of Santiago, covering 1700 km^2 with 550 km of buried cable. The network initially served 250 subscribers, and is gradually expanding to serve 1000 farmers. The $4200 cost per subscriber was paid entirely by local livestock breeders and crop growers.

In 1983 FCh helped install over 600 km of underground cable in external networks of Llanquihue and Osorno. FCh assisted the Telecommunications Undersecretary to unify regulations of telecommunication standards and equipment and participate in a telephone service rate policy study.

Due to the FCh initiative, Chile became the first country in South America to incorporate subscriber funded systems into rural telephony.

GOING BEYOND CHILE

In 1984 FCh was awarded a USAID contract of $121,700 for development of an action program aimed at assisting El Salvador substitute importation of selected vegetables, as well as develop exports of early fresh produce to North American markets. (The terms of reference for the first major FCh assignment outside of Chile appear in Appendix 3.) FCh plans to actively help implement the proposed project, which involves transfer and adaptation of technologies related to production, harvesting, handling, storage, packaging, transport, distribution, and marketing of selected vegetables.

In years to come, FCh intends to do more of this type of outreach activities beyond the borders of Chile.

NOTES

1. Manuel Achurra L. and Emilio Bruggeman A. Mercado Sud-Africano para Productos Pesqueros Chilenos (South African Market for Chilean Fishery Products). Santiago: PROCHILE, October 1984, p. 102.
2. Computerized Process Control Systems (CORFO). Chile Economic Report. February 1986, p.3.

4

Transfer of Marketing Technologies

The most effective marketing departments in service businesses are not necessarily those that are good in marketing... (but) those that are good in marketing... (but) those that get everyone else in the organization to practice marketing.

Managers have three ways to accomplish this: (a) educate employees about the concept, purposes, and applications of marketing; (b) make it easier for employees to be marketers through effective sales training programs and sales aids; and (c) push marketing responsibility down into the organization.

In short, within service companies, the most successful staff marketing directors are those who transform marketing into a line function.

Leonard Berry, President, 1985-87
American Marketing Association [1]

ORIGINS OF MARKETING FUNCTION

FCh is being professionally managed, like any commercial enterprise would need to be managed if it is to survive. The staff at all levels is highly marketing oriented. Indeed most senior professionals are their own best sales persons. That is one of the most outstanding attributes, which sharply distinguishes FCh from many other sister institutions. It was not always that way. [2] Indeed it took all sorts of ups and downs and ups again before a hard-nosed marketing spirit penetrated the entire FCh organization.

Let us first review the shaky beginnings of what has become the FCh Marketing Department (MD), enumerate the jobs it is performing, and take a peak into the future.

The six members of the so-called Landing Party of 1976, charged with setting up FCh, were all marketers. They had to sell the intangible idea of FCh and its purposes to local authorities and potential clients.

In mid-1977, James Parkhill, marketing executive from the Green Giant Corporation, a major US food processing company, was brought to Chile to define what sales and marketing functions FCh should perform and how. Mr. Parkhill's patience ran out by 1979, when he returned to United States. His replacement proved ineffective.

In the interregnum Dr. Roberto Echeverría, the newly contracted former World Bank economist, looked after marketing in addition to his responsibilities for project research and development. The marketing work gradually grew in scope as well complexity: (a) market surveys were being performed as part of pre-feasibility and feasibility studies for FCh projects; (b) products from FCh's technology oriented enterprises .were being sold; (c) marketing services were being carried out on contract to clients; and (d) FCh marketed its own capabilities via seminars, public relations, training, community services, and publications.

In 1983 FCh decided that this crucial and multifaceted work needed to be consolidated under leadership of a capable and imaginative executive experienced in domestic as well as international marketing. An international talent search identified Albin Adam Eng, a Swiss professional with 13 years of European and Latin American experience in the Nestlé Corporation, a universally recognized master marketer of branded food products. The new Marketing Department initiated operations in 1983. A few examples will illustrate how it functions in transferring marketing technology to Chile.

Marketing of FCh Products and Services

The Marketing Department does not sell products and services of FCh enterprises; rather it advises on marketing strategy, helps with preparation of training programs and promotion materials, and handles FCh's public relations. Individual FCh enterprises have their own sales people and are responsible for selling their own products.

From time to time the borders between sales promotion, advertising, and public relations become blurred. The case of the marketing campaign for fresh fish is a case in point. In 1984 the Marketing Department incorporated Don Pez into an overall campaign aimed at promoting seafood consumption via recipes that demonstrate to homemakers how to use fish effectively. This work has been so well accepted that the November 13, 1984, issue of Paula, a leading women's magazine, featured a collection of FCh recipes for Don Pez fresh fish, Chevrita goat cheese, PROCARNE beef, Tongoy oysters and Antartica salmon. The attractively illustrated collection was reprinted in booklet from under the title Fundación Chile: A Guide to Chilean Taste, a Guaranty for International Gastronomy.

In 1984 the Chilean Gastronomy Association (ACHIGA) honored FCh for its efforts aimed at increasing the level and quality of Chilean food; elevating standards of technology required by modern life and nutrition; and increasing awareness of Chileans how to eat better.

In helping to introduce FCh services to new clients, the Marketing Department tends to rely on the demonstration effect. Cynics refer to the strategy as "monkey see, monkey do." Introduction of the Quality Certification Program (QCP), which aims to help Chilean produce exporters to expand into overseas markets and get better prices, is a case in point. In order to find the "path of least resistance" FCh initially approached the trinity of best known, progressive, privately-owned produce packers and exporters: Curto, Standard Trading, and Frupac.

The three companies operate highly automated, fully instrumented, state-of-the-art hydrocooling and packing plants, including cold and refrigerated storage facilities. They were ideal targets for effective use of FCh's QCP services.

In short, the Marketing Department is basically a facilitator helping set up sales systems for FCh products and services.

Marketing Research Capabilities

FCh is equipped to plan, supervise, and interpret marketing research required for pre-feasibility and feasibility studies for FCh own projects as well as do contract work for outside clients. The Marketing Department develops and pretests questionnaires, draws samples, instructs and supervises field interviewers, edits results, tabulates and interprets data, and writes and publicizes resulting reports.

Since the early 1980s FCh has been increasingly able to take on special studies for outside parties. A good example of this type of work is the inventory of the refrigerated transport and warehousing infrastructure. An unbroken "refrigerated chain" is indispensable for protecting the quality of fresh export fruit and vegetables on their way from field to overseas consumers. As a guide to its export promotion work, FCh needed to know what the status of these facilities was. Why were fruits being air freighted overseas, an exceedingly costly form of transportation, if much less expensive maritime freight was theoretically available?

There were no reliable statistics on hand to answer the question. Consequently, in 1984 FCh's Fruit and Vegetable Division undertook a countrywide survey aimed at ascertaining the type, size, and location of existing facilities, and identifying weak links in the cold chain.

The Marketing Department assisted in planning and implementation of the survey. Data were obtained on 867 refrigerated warehouses or chambers with the capacity to accommodate 13 million boxes of apples. On basis of those data an investment program for expansion of the system was drafted. Rather than "armchair invent" needed technologies, FCh staff simply drew on the experience of David del Curto, one of the largest Chilean exporters of fruit to the eastern United States and Canada. On January 18, 1985, del Curto inaugurated, at the port of Coquimbo, a latest-technology cold storage and warehousing center intended for handling table grapes from northern Chile. Initially the facility had an annual capacity for packing 150,000 cartons on pallets. The cost was U.S.$1 million.

When FCh examined the maritime transport used by del Curto, need for major technological changes was identified. Each season del Curto chartered ships for over 70 voyages, half of them destined for the United States and one-quarter to Europe as well as to the Middle East and Far East. In 1984 shipments totalled 7.5 million cartons of fruit, including grapes, peaches, nectarines, cherries, plums, apricots, and asparagus. But the ships did not control the atmosphere in refrigerated holds. Therefore the fruit could not be matured while on its way to markets, so that the quality of the fruit tended to deteriorate, and much of it spoiled altogether. Those circumstances indicated to FCh an opportunity for transfer of controlled atmosphere technology to Chilean reefer ships carrying fruits and vegetables to overseas markets.

A test program was outlined. A multilateral agreement was negotiated with INDURA (an industrial concern) and Delta Lines, for testing the feasibility controlling maturation of fresh produce in overseas maritime transport. David del Curto participated in the pilot project, which consisted of four stages: (a) ascertain behavior of selected produce under currently used maritime container shipping technology; (b) outline improvements of the process; (c) carry out on-land tests in controlled atmosphere simulating a long maritime voyage; and (d) make pilot shipments on basis of gained experience.

The new technology is to generate potentially large benefits: (a) reduce seasonal peak demands for costly airfreight; (b) make it possible to expand access to faraway markets of Southeast Asia such as Hong Kong, Singapore, and Japan; (c) reduce prices; and (d) make Chilean produce more competitive with such Southern Hemisphere producers as Australia, New Zealand, and South Africa.

There was obviously nothing new in the hibernation technology. For over 20 years it has been routinely used in Europe and

North America for extending storage life of bananas, tomatoes, apples, pears, grapes, and other produce that can be harvested prematurely and gradually brought to desired maturity in storage on land or in-transport. But hibernation was a new technology for Chile. That is why FCh became actively involved.

Marketing FCh

Far less visible than Don Pez or PROCARNE products promotion is the assistance provided by the Marketing Department to the rest of FCh staff in their many and varied activities aimed at selling FCh to different publics via such media as seminars, publications, public relations, or community service.

The sophisticated marketing of FCh, which calls for co-opting of public leaders and opinion makers, is an important precedent setting activity.

Printed Media

The first fully integrated booklet aimed at "selling FCh" was based on FCh's first 5 years' operational experience. It focused on basic benefits FCh offered its potential clients. That basic message remains valid. Variations on the hard hitting profit generating theme has therefore appeared in most other publications of the Marketing Department.

FCh markets its story to interested public by profuse use of printed media. FCh professionals prepare bulletins, translate technical texts, and make literature available on such subjects as asparagus and plum growing, apple pruning, growth regulators, nutrient and pesticide treatments with a drip irrigation system, and so on. Appendix 3 lists references to articles about FCh published outside of Chile; publications edited by FCh; and promotional folders. In 1984 alone technical publications, with a total circulation of 38,000 copies, have been distributed to growers, exporters, and professionals.

The Marketing Department drafts press releases, basing them on solidly researched evidence. Each press release outlines what substantial problem is to be tackled by the project, how it is to be done and when and what benefits are expected to accrue to whom. It is all very simple, clear, straightforward, and professional.

In addition to attractively illustrated annual reports, the only other periodic publication of FCh was the 1983-initiated Agriculture Information System (Informativo Agro-Económico, IAE), with two issues of the Bulletin published annually. IAE, the first of its kind in Chile, is to provide agricultural entrepreneurs with reliable and timely agro-economic information needed for making production and marketing decisions at the individual farm level. The response by recipients to this effort surpassed all expectations with IAE advertising covering part of the costs.

Due to its excellent public relations work FCh is receiving a generous amount of coverage by national and local news media. The FCh collection of press clippings for the six months January 1 to June 30, 1985, alone contained over 200 items!

Training

Chile has a number of excellent universities that offer academic courses and curricula in agriculture, food technology, and nutrition. In 1977 Chile had at least 20 food technology related institutions with staff of over 900.

Table 4.1 Staff of Major Food Technology Institutes in Chile, 1977

Type of Staff	Institute of Fisheries Development (IFOP)	Other	Total	IFOP as % of Total
1. Professional	92	318	410	23%
2. Technical	30	129	159	18
3. Administrative	211	127	338	65
Total	333	574	907	36%

Yet there was a shortage of continuity in education, as well as in vocationally oriented, on-the-job and in-house training aimed at improving skills needed for managing and operating special technical organizations as well as industrial and business enterprises. Consequently FCh, as a crucial tool for dissemination of new technologies, runs its own program of workshops, symposia, courses, and seminars.

In 1984 alone FCh offered 39 seminars covering the following topics.

Table 4.2 Seminars Offered by FCh, 1984

Topics	Number
1. Food plant sanitation	22
2. Fruits and vegetables	9
3. Forestry	4
4. Livestock and dairy	3
5. Food processing	1
Total	39

Foreign experts, brought in for specific assignments—such as food sanitation, quality controls, ulex erradication, fruit irradiation, wine and cheese production—usually culminate their stay with a public lecture, workshop, or seminar.

One of the food industry sanitation workshop involved Chilean scientists, FCh technologists, and food quality control experts from North American universities and the U.S. Food and Drug Administration (FDA). Among participants were a large attendance of businessmen from the local food industry. Strict sanitation was being "sold" as a profitable investment, rather than as an expense. Visits to food plants followed, making it possible for FCh to develop a series of instructional slides for use in on-site training lectures all over Chile.

The fruits and vegetable seminars covered such topics as post-harvest handling, preservation and transportation of table grapes, asparagus production, irradiation of fresh fruit, and export potential of fresh produce to European market. Seminars in livestock and dairy products dealt with production technologies for cheese and meat. A special workshop on production of cheese was organized in conjunction with the United Nation's Food and Agriculture Organization (FAO).

The forestry workshops dealt with pulp and paper manufacturing; systems of wood classification; intensive use of wood in housing construction; and control of forestry pests, such as ulex.

The Marketing Department helps publicize FCh events by press releases and media conferences. Seminars thus represent a tool for targeted dissemination of technological data, as well as a means for creating a favorable image for FCh and to sell FCh services.

Public Service

As the good reputation and prestige of FCh spreads, its staff members are increasingly being called upon to serve the "pro bono público." This in turn has reinforced the favorable image of FCh.

A typical case in point was the 1984 appointment of Wayne M. Sandvig, General Manager of FCh, to the newly created National Rice Commission (NRC). The NRC mandate includes change in varieties, harvesting methods, and removal of barriers that might hinder effective marketing of exportable surpluses, which were expected to occur by the late 1980s. Presided over by the Ministry of Agriculture, NRC consisted of representatives from agricultural faculties at the Catholic and Chile Universities, Agricultural Research Institute, National Agricultural Society, Association of Rice Millers, a prominent grower, and FCh.

In 1984 Mr. Sandvig was elected to the presidency of the Chamber of Commerce of the United States in Chile, or AM-CHAM as it is often referred to. He also served as Vice-President of

Chile-North American Institute during 1983-85, was member of the
Advisory Committee of International Executive Service Corps
(IEASC) during 1981-85, and President of Chile Club in 1985.

Similarly, many FCh professionals also participated in such
outside activities as lecturers, adjunct professors, and officers
of numerous institutions of higher learning. FCh case studies
are frequently being used as instructional material, making
educators aware of purposes and activities of FCh.

FCh Invitationals

Use of mass communication media, specific purpose seminars
and workshops, and participation in trade and professional
meetings and conferences are being complemented by highly
targeted outreach toward local and foreign decision makers
related to technology transfer—entrepreneurs, academicians,
politicians, military, journalists, diplomats, professionals,
members of civic associations—who need to know what the FCh is
doing and establish a personal acquaintance with some of its key
staff. The means for achieving this objetive are invitational
round-table lunches, held in FCh's attractive executive dining
room. The events have become little publicized but highly
targeted PR devices of FCh. During 1984 some 37 such events were
held, attended by 334 individuals, including 18 consultants of
foreign countries in Chile.

About two-thirds of visitors belonged to special interest
groups: wives of diplomats and the American Women's association,
the armed forces, and representatives of bi-national chambers of
commerce. Diplomats—ambassadors and consuls accredited to
Chile—were frequently brought in as a means for broadening
international outreach, including search for consultants and
technologies, as well as offering FCh capabilities for potential
assignments abroad.

International Outreach

Success of the catalytic functions that FCh perform, depends
largely on quality of technology sources and the ease of access
to them. FCh is energetically expanding, caring for and feeding
a search network aimed at providing a cost effective means for
locating technologies and sources of assistance required by
clients.

In addition to the enormous pool of talent available from
within the worldwide ITT network, FCh has been particularly
successful at gaining access to smaller European and U.S.
suppliers. Such small firms often have technologies on hand that
are more readily adaptable to needs of small Chilean firms than
what the giant U.S. and West European firms can offer.

Table 4.3 Attendance at FCh Invitationals, 1984

Affiliation	Chilean	Foreign	Total	% of Total
1. Universities	2	–	2	0.6%
2. Research and Development Institutions	3	–	3	0.9
3. Government Agencies and Enterprise	15	–	15	4.5
4. Media	2	–	2	0.6
5. Business and Industry Executives	45	–	45	13.5
6. Professional and Trade Associations, including 26 binational Chambers of Commerce	36		36	10.8
7. Diplomats	2	46	48	14.5
8. Bankers, including one from the World Bank and one from the Inter-American Development Bank (IDB)	–	10	10	3.0
9. Special Interest Groups, including 25 members of Diplomatic Women Association and 80 members of American Women's Association	–	115	115	34.4
10. Armed Forces, including 50 students of the Chilean War College	47	–	57	17.2
Total	162	170	334	100.0%
% of Total	49%	51%	100%	–

Initially, FCh maintained a small office in Darien (Connecticut) with links to North America and Europe. By using normal data bases and technology search aids, this link helped reduce search costs, while providing a mechanism for assisting smaller U.S. suppliers evaluate opportunities for foreign licensing or technology transfer. Work on residual pesticides is an example of the international technology search. Chilean farms have been spraying fruits and vegetables with pesticides in such intensity that residues exceeded guidelines for export to other country markets. In the United States the FDA was turning down Chilean produce at ports of entry. FCh wanted to develop a capability within its existing laboratory to test for residual pesticide on fruits and vegetables destined for export. The search system identified several U.S. testing laboratories qualified and interested to assist FCh to acquire the needed technology and develop required laboratory procedures. The search was completed within two weeks at a cost of under $600.

The continuing presence of an FCh representative in the United States facilitates follow-up communication and assistance to suppliers enabling them to do a more effective job of the transfer of technology and corresponding technical assistance.

FCh's image as an advanced technological center has been strengthened abroad as well. As a result FCh is increasingly being consulted by organizations outside of Chile. A steady stream of invitations are being received for FCh participation in forums that analyze international trade and technology transfer experiences.

In 1983 FCh participated in two events held in the United States: (a) a technology transfer symposium, organized by the American Medical Association (AMA); and (b) a session on countries exporting fresh products to the North American market, included in the annual convention of Produce Marketing Association (PMA), the main trade group of fresh produce wholesalers in North America.

The more than 100 consultants who have been contracted by FCh over the 1976-85 decade constitute a unique, growing and worldwide outreach mechanism. These "FCh alumni" get to know FCh policies, procedures, and activities. They therefore facilitate identification and selection of talents required for performing specific technology transfer functions required by FCh.

In late 1984 FCh signed its first contract with the El Salvador Development Foundation (Fundación Salvadoreña para el Desarrollo, FUSADE). FCh committed itself to prepare an agricultural diversification program by helping identify production opportunities for fruits and vegetables that might either be exportable or substitute for importations. The assignment was basically a variation on the themes pursued on Chilean fruits and vegetables. In February 1985 the job was successfully completed by two FCh staffers, constituting another milestone of FCh marketing its capabilities with a Central American sister institution.

On February 26, 1985, FCh hosted a delegation of executives from a Uruguayan grape and wine cooperative (Cooperativa Agropecuaria de Vitivinicultores del Norte, CALVINOR, in Bella Union) who wanted to utilize Chilean experience gained in quality control of table grapes for exports. FCh arranged visits to packing plants of Siete Amigos, Acomex, and David del Curto.

In 1985 a staff member of the new National Food Technology Center (Centro Nacional de Pesquisa em Tecnologia Agroindustrial de Alimentos, CTAA)—located some 15 miles south of Rio de Janeiro—came to FCh for a short study tour.

CHILEAN SOCIETY OF TECHNOLOGY FOR DEVELOPMENT (SOTEC)

Background

In the early 1980s FCh management increasingly felt that it could not singlehandedly influence the professional, political, economic, and social environment in a way that would systematically facilitate transfer of technology to Chile, and make it an integral part of the country's development process.

During April 1 and 2, 1982, FCh hosted an international seminar aimed at examining experience with implementation of technology development policies in countries with economies similar to that of Chile. The keynote speaker was Major General Manuel Pinochet S., president of FCh, who presented a basic paper on "Technology within the National Development Strategy" (La Tecnología en una Estrategia Nacional de Desarrollo).

The two days of debates that followed this keynote speech clearly indicated that: (a) no country could effectively acquire and absorb technology without having knowledgeable people within the country who know how to buy relevant technology abroad; (b) training an adequate number of these knowledgeable people would require more Chilean professionals to dedicate themselves to R&D work in productive sectors of the economy; (c) the Chilean government would need to assume initial responsibility for construction of scientific and technological infrastructure, putting it at the disposal of the private sector; and (d) the government should establish a technology development policy to guide and provide incentives for the private sector aimed at achieving optimum utilization of R&D infrastructure.

During June 21-23, 1983, FCh brought to Chile Prof. Edward B. Roberts, Director of the Technology Management Program at the Massachusetts Institute of Technology (MIT). He gave a series of lectures for a selected number of Chilean top executives from private and public enterprises and agencies. Dr. Roberts reviewed alternative ways in which Chile could set goals for technology development and provide suitable mechanisms for its effective implementation.

In the wake of the Roberts lectures an ad hoc group of professionals, hosted by FCh, explored the feasibility of establishing a broad-based professional association that would assist in formulation and implementation of technology policies. Their efforts culminated on December 6, 1983, when the Ministry of Justice issued decree 1220, legally establishing the Chilean Society of Technology for Development (SOTEC). During its initial two years, FCh provided SOTEC with office and meetings space and seconded services of Joaquin Cordua S. (a member of the FCh Board Directors, named by the Government of Chile) to serve as its Executive Secretary.

Objectives

The statutes of SOTEC are a case of transfer of organizational technology; they were adapted from statutes of the eminently successful Chilean Biology Society. Specifically, the overall objective of SOTEC is to promote better utilization of technology in the economic, social, and cultural development of the country.

In order to achieve that objective SOTEC is charged with performing the following seven major functions: (a) contribute to strengthening national awareness of technology and understanding of its importance for economic activities, social development, and national security; (b) stimulate better cooperation in matters related to creation and transfer of technology between academia, productive sectors, and government; (c) support studies and research about technological development in Chile; (d) promote improvements of methods for technology development and transfer in research institutions, production enterprises, and government agencies; (e) act as spokesperson for technical opinions of specialists in technology; (f) maintain relations with sister organizations abroad; and (g) execute all activities and contracts that contribute to implementation of the societies objectives.

Activities

In its first two years in existence SOTEC established an impressive track record of activities. SOTEC initially focused on three tasks: (1) draft a paper on the role of technology in development of Chile during the 1985-94 decade; (2) prepare case studies of progressive enterprises that warranted public recognition for their imaginative transfers of technology; and (3) organize seminars for purposes of analyzing projects that have significantly contributed to technological development of Chile.

In 1984 SOTEC initiated six work groups for the purpose of exploring specific topics: (a) defining strategy for national technological development; (b) use of technology for employment creation; (c) establishing a clearing house for case studies of technological innovations; (d) preparing an agenda for a conference on the role of technology in national developments; (e) outlining a communications program for SOTEC; and (f) preparing agendas for future meetings.

During 1984 SOTEC-sponsored activities included: (a) a visit of Dr. Edward Knapp, Director of the National Science Foundation in Washington, D.C.; (b) a workshop on the Colbin-Machicura hydroplant of the National Electricity Corporation (ENDESA); (c) review of outlook for biotechnology development in Latin America; (d) identification of technology requirements for development of the Chilean fruit sector; (e) a review of the technology policy of Israel; and (f) guided visits to local technology-related institutes, including FCh, an activity carried out in cooperation with the Chilean Academy of Sciences.

The 1985 agenda focused on exploration of the following topics: (a) user-oriented information systems about ongoing technological research in Chile; (b) presentations of Chilean case material on successful technology innovators; (c) role of multinational enterprises (MNEs) as a mechanism for technology transfer to Chile; (d) technology education in Chile; (e) role of technology in regional development in Chile; (f) planning technology development of Chile; (g) technology and employment; (h) technologies for reduction of environment pollution in Chile; and (i) ways to increase participation in SOTEC activities by professionals from Chilean universities, technological institutions, and technical agencies.

SOTEC also arranged university-business meetings aimed at exploring possibilities of joint ventures in biotechnology.

Growth of Membership

The initial SOTEC activities were apparently well received by Chilean professionals, academicians, and the business community. By March 1985 SOTEC had 224 members from the following professions.

Table 4.4 Membership of Chilean Society of Technology for Development (SOTEC), 1985

Type	Number	% of Total
1. Business and Industry	74	33%
2. Universities	71	32
3. Government Enterprises and Agencies	51	22
4. Research and Development Institutes	28	13
Total	224	100%

Twelve FCh staffers represented 5 percent of the total SOTEC membership. In 1985 local chapters of SOTEC were set up in Regions 2, 4, 5, 8, and 9. In 1986 membership passed the 450 mark.

In 1986, in accordance with the initial agreement, SOTEC cut its umbilical cords with FCh and set up its own office. FCh's objective of establishing a broad-based lobby for technology development appeared to have been successfully accomplished. SOTEC recognized the great assistance to its creation by bestowing on FCh the 1985 Award for Technological Innovation. The citation praises FCh for "its contribution to promotion of a mentality disposed to use of technological tools as a means of improving the quality of life of the community."

NOTES

1. Service Marketers, Marketing News, December 6, 1985, p. 1.

2.   The field of marketing technology transfer has so far been subjected to benign neglect by both technology and marketing people. The only piece of writing that I have been able to find is a 1981 article in the Harvard Business Review by David Ford and Chris Ryan on "Taking Technology to Market." It has a small section called Dangers of Technology Transfers. See: The Marketing Renaissance, edited by David E. Gumpert. New York: John Wiley, 1985, pp. 326-27.

3.   In 1985 David del Curto, with plants in Malloco and Coquimbo, exported 9 million boxes, or about 20 percent of the country's total produce exports. Curto has 500 permanent employees, and hires up to 5000 seasonal workers for crop picking.
     Standard Trading Company (STC), a Castle and Cooke associate, started operating in Chile in 1981. In 4 short years it has become the second largest produce exporter in the country. Its plants in San Francisco de Curimón, San Bernardo, and Colchagua Copiapó are staffed by some 200 permanent employees. Castle and Cooke markets its products under the well-known Dole brand. STC has effectively used its parent company's vast marketing, technological, and financial resources.
     FRUPAC owns packing plants in San Felipe, Santiago, and Rancagua, and covers the North American market through its wholesale operation in Philadelphia. Frupac joined hands with Britain's Mack & Edward in getting established in West European and Middle Eastern and Far Eastern Markets.

5

Financing

Management of the portfolio of FCh's liquid assets
has permitted us to increase financial returns at
rates consistently and significantly higher than
average rates yielded by investments in the
Chilean capital market. Annual incomes from
operations have been steadily increasing in a way
that practically assures future financing at
current levels of activities. Furthermore,
investments already made in subsidiary and
affiliated enterprises are expected to shortly
turn the corner and yield substantial cash
benefits.

M. Wayne Sandvig
Report to FCh Board of Directors XII/17/1985

During the first decade of its activities, FCh has carefully
husbanded the U.S.$50 million seed capital it received from its
parents ($25 million from the Government of Chile and $25 million
from ITT). The funds were only gradually released as FCh proved
that it could effectively use the money to finance the ever
growing operations and/or prudently invest it in the local
capital market. During the first period 1976-78, $24 million of
seed money was made available to FCh, while the remaining $26
million was released at an average of $4 million annually during
the 7-year period 1979-85.

Within its first decade of activities, FCh invested about
half of its seed money into building its headquarters and
associated enterprises, as well as subsidizing annual operations.

On January 13, 1983—a warm summer day—a man-made financial
earthquake hit Chile. Because of their excessive indebtness the
government took over 5 banks, including two of the largest

private banks. The two largest conglomerates collapsed. Thousands of holders of related mutual fund shares and debentures were instantly impoverished. Only 70 percent of time deposits in the two of the intervened banks could be recovered. Until 1981 FCh's portfolio included many of these papers. In late 1982 investments were transferred to more solid bank and government papers. Consequently FCh came out unharmed by the financial crisis.

As of October 31, 1985, over 80 percent of FCh's liquid funds were conservatively invested in short-term paper in 8 leading banks:

Table 5.1 Short-Term Investment Papers in FCh Portfolios, October 31, 1985

Bank	Millions of 1985 pesos Amount	% of Total
1. Central Bank of Chile	936	23%
2. State Bank	621	15
3. Citibank (New York)	440	11
4. Bank of Chile	328	8
5. City Bank	323	8
6. Centro Bank	285	7
7. American Express	240	5
8. Banco Sud-Americano	185	4
9. Sub-total	3,358	81
10. Other 10 institutions [a]	789	19
11. Grand Total	4,147	100%

[a] Includes: Banco de Crédito, Hongkong & Shanghai Bank, Banco O'Higgins, Banco de Santiago, Banco Español-Chile, Banco Industrial Com-ex, Banco Urquijo de Chile, Chicago Continental, Banco del Trabajo, and Bank of America.

This conservative investment portfolio produced a yield consistently and substantially some 3-4 percentage points above the average rates in the overall Chilean capital market.

From its inception FCh was mandated to gradually increase its degree of self-financing. A target of 50 percent was to be achieved by about 1990. In the period 1980-85 FCh was well on its way toward achieving that objective, in that sales, expressed as percentage of cost of operations, increased from 12 percent in 1980 to 39 percent in 1985. The amounts in Ch$ were as follows:

Table 5.2 Degree of Self-financing FCh Operations, 1980–85 (in 1985 Ch$ millions)

Year	Sales	Cost of Operation	Sales as % of cost of Operations
1980	Ch$ 71	Ch$ 595	12%
81	99	621	16
82	105	553	19
83	199	686	29
84	250	645	31
85 [a]	238	619	38
Total	Ch$962	Ch$3,719	26%
U.S.$ million equivalent [b]	$ 5.7	$ 21.2	

Source: Chilean Foundation, Annual Report 1985

[a] See next paragraph.
[b] Exchange rate Ch$ 180 to one U.S.$.

Table 5.3 Income from Sales and Expenditures, FCh, January–October, 1985

Item	Ch$ million	% of Total
1. Sales Expenditures for FCh products		35
2. Operations		492
2.1 Food Technology	207	
2.2 Forestry	58	
2.3 Marketing	50	
2.4 Electronics and Telecommunications	34	
2.5 Finance and Administration	61	
2.6 General Management	51	
2.7 Unallocated Overhead	31	
3. Earthquake Emergency		28
4. Losses from Associated Enterprises		64
5. Total Costs		619
6. Income from Sales		238
6.1 Products	43	
6.2 Services	191	
7. Income from Sales as % total Costs		38%

During the 10-month period from January 1 to October 31, 1985, the income from sales of FCh goods and services covered 38 percent of total costs.

During the first 10 months of 1985 the Ch$ 50 million costs of marketing consisted of the components shown in Table 5.4 (in Ch$ millions):

Table 5.4   Cost of Marketing, FCh, January-October, 1985

Component	Amount	% of Total
1. Personnel	25.6	51%
2. Public Relations	6.7	13
3. Travel	6.5	13
4. Prorated Joint Costs	2.5	5
5. Miscellaneous	9.0	18
Total	50.3	100%

In years to come, as income from sales of FCh goods and services increasingly pays for a large proportion of current costs, FCh plans to: (a) diversify its conservative investment portfolio; (b) mobilize more funds from multinational agencies and governments; (c) take on social projects, for which national or regional governments allocate adequate counterpart funds from regular budgets; (d) submit tenders for projects eligible under government incentive payments for technological development; and (e) seek out financial support from other institutions willing to become shareholders in FCh.

6

Project Monitoring and Performance Evaluation

Research and development funding is not a cure-all for technology... (transfer to developing countries). Effective management and deployment of technological resources are more important than the numbers of dollars invested.

National Research Council
News Report, April 1986

FCh was conceived as a business, not a charitable organization. To justify its continuous existence its activities therefore have to generate long-term profits for FCh, its clients, and the Chilean society as a whole.

Guiding Principles

Transfer of technology is an innovative and risky business. Inevitably there will be failures no matter how carefully individual projects and activities have been identified, selected, prepared, ex-ante evaluated, and executed. The challenge is to weed out nonperformers early in the game, and to lend a steady hand to performers. To do that FCh installed a monitoring system called Annual Evaluation of Results of the Operations Program, better known by its AEROP acronym.

AEROP is being prepared by an outside consultant, who directly reports to the Board of Directors rather than the Director General of FCh. This way he maintains a detachment from the pressures of day-to-day operations.

The evaluation mechanisms and quantification criteria used by FCh aim at reaching two objectives: (a) estimate the private and social benefits expected to be generated by adaptation of technological improvements; and (b) quantify the proportion of total benefits that can be directly attributed to actions, of FCh.

Expected benefits include: (a) income of FCh, producers and state (via income tax); (b) social value of foreign exchange earned and employment created; and (c) consumer surplus, whenever it can be quantified.

AEROP is a courageous attempt at running FCh affairs in a business-like manner. The evaluation methodology is under continuous scrutiny to make it more relevant, accurate, and a timely tool of management. That purpose is unassailable, and procedures pursued worth emulating. For this reason there is no need to go into long litanies about "how problematic all this is." Neither would it serve any useful purpose to criticize the procedures developed. The reader will need to use own judgment on these matters.

It seems more constructive, for purely illustrative purposes, to present the data "as is," together with enumeration of assumptions, including explanatory footnotes. Cost and income values were converted into U.S. dollars at the December 1983 exchange rate of 87.1 pesos to 1 U.S.$.

This chapter consists of two main parts: (a) methodological considerations guiding the cost/benefit measurements of FCh activities; and (b) review of major results based on the 1983 AEROP.

Measurement Methods

FCh engages in three major types of technology transfers: (a) technical assistance in development of products, processes, and technical improvements, aimed at boosting the overall efficiency of the enterprise; (b) development of new products containing innovative technological components; and (c) demonstration enterprises based on new technology.

Every technology transfer involves at least three participants: (a) FCh, as transfer agent; (b) buyers and users of the technology, who contract with FCh for services related to transfer of the technology; (c) The rest of the economy, which consists of five types of beneficiaries: emulators, who increase their productivity by copying the technology from innovators; consumers, who buy products at lower prices due to passed-on savings in costs brought about by the newly introduced technology; beneficiaries from public services funded by additional income taxes collected from users of the transferred technology; the individual, who obtains remunerative employment generated by the technology; and the national treasury, which improves balance of payments either through efficient substitution of importations or increase of exports.

In measuring cost/benefits, FCh attempts to quantify magnitudes that can be readily ascertained, such as FCh's own outlays and revenues generated in transactions with its clients, plus some of the evident social benefits derived from diffusion of technologies beyond initial clients.

Social benefits are not readily quantified. In most cases FCh therefore merely enumerates them. Yet, it is evident that in the long-run, indirect benefits tend to be greater than direct private benefits derived by FCh and its clients.

AEROP monitors ongoing activities so as to assist policymaking and review by the Board of Directors of FCh and provides clues to macro-economic resource allocation based on long-run impact of alternative technology transfers.

It takes years before the impact of a technology works itself out over the "natural life" of the innovation. AEROP established a workable time frame for the 1983 AEROP, which is to serve both objectives. It covers fours years: (a) comparison of actual and budgeted results for 1982 and 1983, and (b) comparison of initially budgeted with revised projections for the 4 year period 1982-85.

This time frame reflects the realization that, even though the cost and benefits of FCh actions cannot be evaluated on basis of one year only, it is feasible, realistic, and desirable to annually monitor activities of each major program. That way the Board as well as the Director General of FCh can be provided with valuable feedback required for initiating prompt remedial actions.

Results

The continuous auditing of cost/benefits is adapted to idiosyncrasies of each ongoing activity. This makes it possible to arrive at widely varying interpretations and proposals for future actions.

Global Assessment

Measurement of direct costs and income is the easier part of the audit. During 1982-85 the costs of the 10 main activities are projected to $10.6 million, direct income to $5.2 million, or about half the cost. The financial self sufficiency varied all the way from a 15 percent low for development of jojoba bean plantations, while microprocessing generated income surplus over costs of 144 percent.

Case of Microprocessors

In 1976 FCh started to develop applications of micropro-cessors. The initial steps focused on demonstrations and training. During the 1982-84 period FCh carried out contracts for commercial use of microprocessor technology adapted to specific tasks in six private and public enterprises.

Table 6.1   Income/Cost Assessment of FCh Activities, 1982-85

| Activity | US$ 1,000 equivalents [a] | | Income as % of Cost |
	Income (1)	Costs (2)	(3)
1. Marine Resources	$ 613	$ 1,543	40%
2. Livestock	844	1,435	59
3. Fruits and Vegetables	1,102	2,113	52
4. Laboratory Services	264	562	47
5. Pilot Plant	452	1,073	42
6. Technology (Jojoba)	19	127	15
7. Forestry	726	1,734	42
8. Microprocessing	661	442	144
9. Marketing	379	424	74
10. Other	187	1,134	21
Total	$ 5,247	$ 10,637	48%

Source: Luis A. Adriasola, Annual Report on Results of Evaluation of the 1983 Operations Program (Informe Anual de Evaluación de los Resultados del Programa Operacional 1983 y su Continuación hasta 1985), Fundación Chile, September 1984, Table 2, p.11.
[a] Converted at 87 pesos per U.S. dollar.

Table 6.2   Microprocessor Contracts, FCh, 1982-84

Year	Project	Client
1982	1. Feasibility of Remote Control of Energy Generating Centers	El Teniente
	2. Feasibility of Load Dispatch	El Teniente
	3. Intake (Bocatoma)	ENAP
	4. Refrigeration (Enfriamiento)	ENAP
	5. Fire Combat Pumps (Bombas Contra Incendio)	ENAP
1983	6. Control System at Disputada	EXXON
	7. Remote Control at Salvador	CODELCO
	8. Communication System at Chuquicamata	CODELCO
1984	9. Automatic Fuel Dispenser for Gasoline Filling Stations	COPEC
	10. Information Recording at Oil Platform	ENAP
	11. Date Recording and Communication Systems for Oil Pipeline	PETROX

The 1982-85 socioeconomic impact of microprocessing was favorable in that FCh income covered 117 percent of variable costs: 59 percent via direct income to FCh and an additional 58 percent via partial benefits for clients.

Three major unquantifiable benefits were being generated: (a) reduction of overall costs for the country due to early introduction of technology brought about by FCh's catalytic action; (b) training of professionals in client enterprises, enabling them to become self-sufficient in applications of the technology; and (c) establishment of local capability at FCh, which in 1983 added up to 10,000 professional hours and saved the country at least $0.5 million in foreign exchange, that otherwise would have to be spent on adapting standard programs to specific uses.

Affiliated Enterprises

Prior to establishment of the five affiliated FCh enterprises (Salmones Antártica, Cultimar, Procarne, Caprilac and Berries La Unión), internal rates of return (IRR) analyses were undertaken together with corresponding financial projections. The form of presentation of the data has been standardized. A good example is the summary tables for the IRR and Net Value Added (NVA) calculations undertaken for PROCARNE under three types of sales assumptions:

Table 6.3 Sales Projections of PROCARNE, 1984-88 and Beyond

Type of Sales Assumptions	Annual Sales Projections (in Ch$ million)				Results for Period 1983-93	
	1984	1986	1987	1988 and beyond	Net Value Added (Ch$millions)	Internal rate of return (%)
1. Pessimistic	286	604	755	794	Ch$ 31	15.3%
2. Realistic	319	673	841	885	191	31.8
3. Optimistic	350	738	923	921	337	46.5

Within the framework of FCh's monitoring system the above data are being annually updated and are actual results compared with initial estimates. Causes for major discrepancies are then identified. This feedback provides managers of individual enterprises with indications of alternative remedial actions.

The 1984 update of total benefits generated by the five affiliated enterprises were estimated to be about Ch$3.1 billion, divided half and half between direct private and social benefits:

Table 6.4 Benefits Generated by FCh Affiliated Enterprises, 1984

| Enterprise | Benefits in Ch$ millions | | | Social benefits as % of total benefits |
	Private (Net Value Added)	Social	Total	
1. Procarne	Ch$ 807	Ch$ 458	Ch$ 1,265	36%
2. CAPRILAC	52	37	89	42
3. CULTIMAR	60	78	138	57
4. Salmones Antártica	507	801	1,308	61
5. Berries La Unión	120	163	283	58
6. Total	Ch$ 1,546	Ch$ 1,537	Ch$ 3,083	50%

Source: M. Wayne Sandvig, Activities of FCh, 1976-85, Report to Board of Directors, December 17, 1985.

In the last 1976 meeting of the FCh Board of Directors one of its members pointed out that "the foundation would have fulfilled its objectives upon achieving sufficient financial flows from its operations to permit its maintenance into the indefinite future." That objective has been achieved within the first decade: FCh generated enough income, from its operations as well as its investment portfolio, to more than cover expenses. Furthermore, FCh provided significant side-benefits to the community: (a) improved productivity, due to effective utilization of technologies transfered to clients by FCh, helped to increase income for producers, lowered prices for consumers, and boosted tax receipts for the government; (b) affiliated enterprises, created by FCh, stimulated economic activity as well as offered new employment opportunities; and (c) as FCh stays in operation in years to come these benefits will continue flowing.

The above examples of the methods and results of the Annual Evaluation of Results of FCh Operations Program (AEROP) are illustrative of the exceptional attention given to developing an evaluation system of potential benefits—private and social—at the start, during, and at completion of projects. Private benefits are being estimated with painstaking care even though it is recognized that FCh can never know exactly how much net profits its clients will generate.

In a letter of August 8, 1985, to members of the FCh Board of Directors, Mr. Sandvig laconically reported: "The evaluation of the 1976-84 action program of the Fundación Chile indicates a social return of at least 24 percent ... This compares favorably to the social discount rate of 12 to 15 percent experienced during the years in which the Fundación operated."

To my knowledge none of the approximately 150 food technology related sister organizations of FCh have an elaborate monitoring system. [1] Therefore they stand to learn a great deal. Fortunately FCh is ready and willing to share its experience with all bona fida data seekers.

NOTES

1. B. F. Buchanan and C. Stewart, Food Technology Resources in Southern Latin American Countries. Washington, D.C.: USAID, May 1977.

7

Achievements of the First Decade

> Under Mr. Sandvig's leadership, Fundación Chile
> increased in size from 17 to 250 employees, and net
> worth from $7.2 million to $31.7 million. It has
> become a self-sustaining organization that delivers
> significant benefits to the Chilean community through
> transfer of technology and creation of new business
> enterprises.
>
> Citation for the Bernardo O'Higgins
> Medal of Honor, October 13, 1986 [1]

GENESIS

From the first modest steps, taken in the mid-1970s, FCh has
come a long way in its work of building bridges for transfer of
new technologies to Chile from abroad.

In retrospect, the creation of FCh in 1976 as a joint
venture between the International Telephone and Telegraph Company
(ITT) and the Government of Chile, was a creative and brilliant
resolution to a bothersome problem: compensation to ITT for
nationalization of Chilteco, its Chilean subsidiary.

In 1974, once the decision was made to set up an organization that would promote transfer of technologies to Chile, the
joint venture partners moved quickly. There was little opportunity for elaborate selection from among ITT's huge arsenal of
staff with international experience: Dr. Steward S. Flaschen,
who was sent to Chile for the initial 1984 negotiations, had
little previous exposure to work outside the United States; Dr.
Robert Cotton, the first Director General of FCh, had none.
Neither spoke Spanish. Yet their unlimited enthusiasm, combined
with intimate knowledge of how to get things done within ITT,
made it possible to put FCh rapidly into orbit.

The cost of getting FCh established was $1.3 million. The net worth (underline)patrimonio(/underline) of FCh was $7.2 million. The operation costs, authorized by the Board of Directors for 1977, were $2.8 million.

In order to establish "early credibility" FCh had to produce some visible results soon after initiation of its work in 1976. A feeding program for the poor, the so-called CONPAN project, was run under auspices of the National Food and Nutrition Council (Consejo Nacional para la Alimentación y Nutrición; the acronym CONPAN means "with bread" in Spanish). Within the framework of CONPAN the Government of Chile (GCh) provided food for needy school children. Early efforts were only partially effective because of serious problems in supplying schools with proper foods, and in maintaining high sanitation standards in food preparation and service in school environments. This situation provided an opportunity for FCh scientists—working in cooperation with Dr. Fernando Monckeberg, head of CONPAN and professor of nutrition at the University of Chile—to develop a wholesome, nutritious, ready-to-eat biscuit, which could be readily manufactured by local bakers.

The CONPAN project was accompanied by a wide variety of other FCh mini-projects. Among the early ones was establishment of an aquaculture center in Coquimbo. It was a controversial project not only because its location, but also due to uncertainty about species that would be suitable for production under Chilean conditions, as well as doubts about size of potential domestic or export markets. Those are the normal headaches of innovators. Next in line were projects such as: (a) improvement of processed fruits and vegetables; (b) utilization of apple rejects; (c) technical assistance to canning plants; (d) study of potato processing; (e) technical assistance to edible and industrial oil industries, aimed at refining fish oil and utilization of subproducts; (f) establishment of kitchen gardens within the framework of the Las Acacias self-help program; and (g) search of remedies for lactose intolerance with experiments performed on prison populations (most results of the study were lost when some of the prisoners escaped from jail). Several of the early projects provided the seeds from which important future FCh activities grew.

FCh helped a firm wishing to produce fruit pulps for Swiss-type yogurts to do engineering designs and product formulations. By 1984 the client was able to supply 80 percent of local consumption, which substituted for most of the previously imported pulp.

FCh technicians designed a plant for production of dietetic rice-flour, which uses discards from rice milling. FCh specified equipment, formulated products, and initiated manufacturing.

During the first couple of years—the infant stage—of the existence of FCh, objectives for most projects were not clearly defined and their technological focus was vague. These shortcomings were due to the strictly exploratory nature of the initial work. FCh was courageously dipping its toes into the sea of potential opportunities.

In 1977, M. Wayne Sandvig took over the position of Director General. His first task was to accelerate the speed by which the long established ITT Management System for Research and Development (R&D) would be adapted to FCh requirements for Systematic Management of Technology Transfers (SMTT). The resulting hybrid system was based on ITT's R&D cases, defined as "groups of activities that are being developed in order to achieve predetermined objectives within annually approved budgets" (i.e., "management by objectives"). SMTT made it possible for FCh to control planning and budgeting more effectively, as well as matters related to personnel administration.

FCh created a system of financial management, including handling of its investment portfolio, as well as developing methodology for evaluation of the socio-economic impact of individual projects. Management of liquid assets generated yields that were habitually, and significantly, 3-4 percent above average yields in the Chilean capital market.

Since 1980 FCh began to undertake annual evaluations of the socioeconomic impact of its technology transfer projects. Although some FCh staff initially viewed these systems with skepticism, they soon came to accept them as useful tools of management control.

TECHNOLOGY-BASED ENTERPRISES

Development of technology-based enterprises proved to be an important contribution to the Chilean economy. Two of these enterprises were set up as subsidiaries of FCh: Cultivos Marinos Tongoy (Cultimar), which focuses on production of Pacific oysters, and Salmones Antartica, for salmon ranching in southern Chile. Three enterprises are affiliated with FCh: Caprilac production of quality goat cheeses, PROCARNE for vacuum packed boxed beef, and Berries La Union (BLU) in Region 10, an integrated agro-industrial complex that is to produce and export a variety of small fruits (blueberries, gooseberries, strawberries, and raspberries).

The initial 2-3 years of pioneering enterprises tended to be periods of shakedown and adjustment. By early 1986 Cultimar was behind schedule by at least one year, primarily due to slow development of markets. Currently the main challenge is the long spawning period. During November to February Chilean oysters either spawn, are getting ready to spawn, or are recovering from it. Production of sperm and eggs weakens the oyster and makes the flesh flimsy, musky, and unattractive to consumers. Treatment of the embryos with cytochalasin B and a dimethylsulfoxide (DMSO) solvent sterilizes the oyster with three sets of chromosomes instead of the normal two. The triploids do not spawn, so that the meat stays fleshy year-round.

For triploiding to work, an oyster egg must be treated with the diluted solution of cytochalasin B and DMSO within 30 minutes of fertilization. Some 15 minutes later the eggs must be washed clean with DMSO-15. Westcott Bay Sea Farms in the State of Washington have developed a way to make ripe oysters "spawn on command." The eggs can then be gathered and treated more readily. The ultimate goal is a tetraploid, an oyster with four sets of chromosomes. Such a mollusc would spawn, but its offspring would be sterile. Champion "bull and cow" molluscs could then produce thousands of triploids each.

The U.S. Food and Drug Administration (FDA) has approved triploid molluscs; European and Japanese oyster farmers are also interested in triploids; Cultimar started to look into the possibility of testing and refining that technology.

In the case of Salmones Antartica, a few fingerlings, released in the late 1970s, actually returned as grown salmon to spawn in their natal waters of southern Chile. The El Niño phenomenon, consisting of untimely warming of the cold and nutrition-laden waters of the Humboldt current, could have been one of the potential causes. Or some more permanent environmental limitations may have militated against ocean ranching technology. Under those uncertain circumstances, FCh initiated a search for alternative technologies such as cage farming (crianza en jaulas). Although more costly to operate than ocean ranching, the cage technology offers far more control over the growth of the fish. Yields and unit costs of both technologies are being compared.

In 1985, salmon cultivation in cages was further expanded with construction of installations at Changagnitad in the Dalcahue Canal. The three facilities of Salmones Antartica—including the Chiloé island and Puerto Chacabuco in Aysén—thus reached a combined annual production capacity of 400 tons. During 1985/86 Salmones Antartica exported 300 tons of fresh Pacific salmon to the United States and Canada, or about 30 percent of total Chilean production of this new species. Salmones Antartica makes its own specially formulated pellet feed, which it also sells to other salmon farms in Region 10. The open sea ranching is being continued at Curaco de Vélez and Astilleros on the island of Chiloé. FCh management "looks at the future of this enterprise with cautious optimism."

Within two years of start-up, the PROCARNE processing plant for vacuum packed boxed beef reached operational break-even point, even though it continued to face the vagaries of the marketplace.

In 1986, after a promising beginning, the Caprillac plant for production of pasteurized quality goat cheeses had to suspend operations. This was simply because the small goat breeders in the foothills of Region 4 could not adequately reduce the huge seasonal fluctuations in supply of raw milk. Operations became too costly with the plant, which was running way below capacity for most of the year.

The newly established Berries La Union (BLU) enterprise has gotten off to a flying start by utilizing advanced cultivation and post-harvest handling technologies. Most of the BLU berries are marketed abroad under the Berry Good brand. All indications point toward a successful development of exports of a wide variety of fresh berries.

ITT has access to and/or owns substantial telephone and telecommunication related technologies, some of which were deemed to have potential for adaptation to Chilean conditions. Consequently in 1976, the year when FCh officially got under way, pre-feasibility studies for rural telephony were initiated. During 1979/85 six subscriber carrier systems were installed in the following locations: La Herrera, Llanquihue, Osorno, Colina, Calera de Tango, and La Ligua. The cables were buried underground because many years of U.S. experience clearly indicated that these installation costs averaged some 45 percent less than open-air cables.

Hand in hand with the early adaptation of rural telephony, FCh also sought transfer of some of ITT's electronics technology. The work aims primarily at facilitating automatization of industrial processes by microcomputerization, and development of standardized integrated control systems adaptable to a wide variety of uses. In the early 1980s FCh was able to create the SIC-15 modular computer system based on adaptation of microprocessor technology. The versatile SIC-15 software is designed to control a wide variety of industrial applications ranging all the way from simple processes to complex operations adapted to specific geographic and functional requirements. The SIC-15 system, which was developed by FCh staff, is comparable to those produced in highly developed countries. Within the profession SIC-15 is considered "to be at the leading edge of technology" (tecnología de punta).

Assembly of the first-class human resources, their in-service training and a huge capital investment are the most valuable assets of the FCh microcomputer unit. SIC-15 evidently has excellent additional potential for wide ranging applications in mining, energy, and all sorts of manufacturing. SIC-15 not only substitutes for importations, saving Chile scarce foreign exchange, but FCh clients are being provided with a permanent in-country source of training and equipment maintenance.

As would be inevitable for any courageous innovator, FCh got involved in a few goose chases that eventually had to be abandoned. The Technical Assistance Center for the canning industry (CAT) demonstrated the inability of FCh to overcome the reluctance of enterprises to jointly invest in long-term industry-wide technology improvements. FCh also put on the back burner development of processes for making lupine seeds edible, and irradiation of smoked salmon.

As pointed out earlier, the Caprilac goat cheese processing plant, which initially looked like a very promising venture, had to be discontinued simply because goat herders in the remote hills of Region 4 could not deliver a reliable year-round supply of raw milk. High overhead would have made it economically unrewarding to operate the plant only during the spring and summer months.

FCh has never brushed under the carpet its unsuccessful activities. Rather, it has analyzed causes of these failures in order to avoid repeating the errors in future activities.

EXPORT PROMOTION

From its very inception FCh has assigned high priority to projects that would favorably influence balance of payments either by increasing exports or substituting imports.

Three of the FCh technology oriented demonstration enterprise (salmon, oyster, berries) are primarily export-oriented. So was the huge technical assistance program aimed at producers of fruits and vegetables (apples, pears, table grapes, and asparagus), which made it possible for Chile to profitably enter the off-season markets in the Northern Hemisphere. The rapidly growing Quality Certification Program (QCP) likewise aims to facilitate nontraditional exports.

FCh has worked hand in hand with the Agricultural Research Institute (INIA), which aggressively promotes production of exportable varieties of crops and related technologies. INIA introduced improved pistachios, nuts, pecans, persimmons, figs, and kiwi fruit. The successful development of improved asparagus varieties by FCh resulted in a fivefold increase in exports during 1982-85.

In retrospect it appears some opportunities were missed. A case in point are the plastic greenhouse or tunnel structures, widely used in the Mediterranean region for growing early vegetables. With the benefit of 20/20 hindsight it would have been useful to include this technology in development of fresh vegetables for markets in the Northern Hemisphere. That technology might make parts of the Chilean crop mature a month or two earlier than open air cultivation, and reach the North American and West European markets at the highest seasonal prices. Asparagus comes readily to mind as a crop that would have been a suitable pilot project.

QUALITY CERTIFICATION PROGRAMS (QCP)

FCh helped to create and firmly establish a program of quality control and certification for export products, based on standards and norms used in importing countries. By 1985 FCh's

QCP centered on fresh and processed foods (red meats, fruits and vegetables, fish and shellfish, and different types of preserves), wine and distilled spirits, processed wood, wood-intensive homes, and other intermediary and finished products. FCh also entered into agreements with sanitation control organizations of countries that traditionally import from Chile. The most recent of these agreements was signed with the South African Bureau of Standards (SABS).

PROCARNE is the first firm to use the vacuum-packed boxed beef method based on technology supplied by FCh, which also helped Chile adopt, for internal use, internationally recognized standards for beef cuts.

The FCh quality control and certification program for fresh fruit and vegetable exports is the largest in the country. The area covered by the program extends all the way from Region 3 to Region 8. Some 10 million boxes were inspected during 1985/86. For wines and grape-derived alcoholic beverages the program covers the entire production process, including raw materials and final products.

FCh also does on-farm pesticide residue control for fresh fruits and vegetables destined for exports.

FORESTRY-RELATED WORK

FCh's work in forestry has focused on management of native stands as well as plantations of radiata pine, nurseries, saw mills, pulp and paper manufacturing, marketing, and utilization of residues and by-products. By introducing and adapting technologies that have already been proven effective in other parts of the world, FCh helped improve productivity in some of the nation's forestry enterprises.

Specifically, during 1985 FCh: (a) promoted utilization of forestry waste and by-products as fuels to generate energy; (b) supervised establishment and administration of tree nurseries that have produced a total of 13.5 million pine trees; (c) helped to improve technology used to dry native woods; (d) evaluated potential and profitability of methods conducive to growing knot-free radiata pine. The majority of these technology transfers have been sponsored by businesses groups, enabling them to generate benefits efficiently and economically.

The FCh Forestry Department is helping leading firms to improve overall competitiveness of Chilean products by using quality as a promotional feature. FCh joined hands with the University of Biobio to prepare classification standards for radiata pine wood and timber. FCh has also developed standards and quality control mechanisms to provide consumers with relevant information. Under auspices of an agreement with the Chilean Wood Corporation (CORMA), FCh is carrying out a technical study to determine a grading system for sawn Radiata pine components for use in construction as well as development of a grading system for nonstructural wood products.

A group of Chilean forestry firms, including plywood and paneling manufacturers, sponsored testing of boards and walls designed for use in wood homes. The purpose was to evaluate behavior of these materials under specified mechanical/structural and physical/environmental conditions. The positive results helped broaden opportunities for use of wood elements and structures in building.

FCh is providing technical services to assist firms and investors to identify, analyze, and implement new forestry projects. FCh thus prepared a feasibility study for construction of a thermoelectric power plant fueled by waste from sawmills, driers, and wood processors, as well as determined the feasibility of building high-quality prefabricated houses for export. Both projects appear to be attractive investment opportunities. Their implementation would have a favorable socioeconomic impact on low-income families in the forested areas of Southern Chile.

Since 1983 FCh has been promoting use of wood in home construction. These efforts led to establishment of an inter-institutional consortium made up by FCh, the University of Chile, and the National Forestry Corporation (CONAF). In 1985 the consortium made advances: 15 energy-thermic homes were built and quality controlled; changes were made in building legislation and codes; thus reducing obstacles to use of wood; technical assistance for manufacturing of various wood components for homes and a quality certification program were provided. Furthermore, via a publicity campaign that included use of printed and television media, FCh helped improve acceptance, by builder of wood as suitable home construction material.

The close relations that FCh staff members established with their forestry counterparts in New Zealand made it possible for FCh to act as one of the catalytic agents in helping to trigger at least three joint ventures. Cholguán Wood Panelings (CWP) and Carter Holt Harvey (CHH) combine Chilean raw material with New Zealand technology. The contemplated $25 million plant is to be located in mid-southern Chile, have capacity of 83,000 m^3, employ 300, and export $25 million annually. Furthermore, by mid-1986 Carter Holt Harvey signed a $164 million with Chile's Angelini conglomerate of fishing, lumber, paper, and dairy industries. The new joint venture, called Inversiones y Desarrollo Los Andes (IDLA) is to include investments in forestry, fisheries as well as in Angelini's controlling stake in the Compañía de Petróleos de Chile (COPEC). IDLA is to incorporate COPEC's 228,000 ha timber land, a pulp mill with annual production capacity of 360,000 tons and a fishing fleet with annual catch of 700,000 tons.

In addition, IDLA will have access to COPEC's service stations, which handle about 43 percent of Chile's gasoline consumption as well as COPEC's 45 percent stake in a coal mine producing 850,000 tons of coal annually, and 2 power distribution company. The attachment of Carter Holt Harvey to the COPEC empire will thus facilitate its access to the world's single largest Radiata pine resource and a world-scale fishing fleet.

FCh also served as one of intermediaries, which brought about a joint venture the Chilean Compañía Manufacturera de Papeles y Cartones (Comapac) and Fletcher Challenge (FC), the largest New Zealand industrial and forest products company. FC operates the country's only newsprint mill and own Crown Forest Industries, a lumber and paper products subsidiary in British Columbia (CFI/BC). According to the agreement, reached in late 1986, FC is to invest $50 million in the contemplated new enterprise, consisting of 36,000 ha (78,000 acres) of mainly radiata pine in the Concepcion area, and a paper mill producing 74,000 tons of newsprint and telephone directory paper for South American markets. With the COMAPAC joint venture Fletcher challenge aims to develop a strong position for Radiata pine on world markets before large volumes of raw material become available from New Zealand forests in the late 1990.

These types of joint ventures, which match foreign technology and capital with Chilean raw materials, represent a fulfillment of one of the key objectives of FCh's existence.

Be that as it may, in 1985 FCh was active in practically every one of the country's twelve administrative regions, ranging all the way from fishing in Region 1 in the North, to conducting experiments in salmon ranching off the coast in Region 12 in the extreme South, as well as working on computer applications to aid petroleum production operations. FCh's contribution to technological development was especially evident in Region 10: (a) new technology was provided for rural telephony; (b) computerized management of dairy farms; (c) pacific oyster and salmon cultivation; (d) conversion of forestry wastes into fuels to be used in generating electricity; (e) evaluating the economic impact of hydro-electric projects; (f) harvesting forests; (g) gathering information to assess economic feasibility of cultivating alternative crops; and (h) introducing small fruits as a means to diversify agriculture" (see Appendix 3).

TECHNOLOGY MARKETING

Marketing is subjected to benign neglect in most technological research institutes of Latin America and the Caribbean. Not so in FCh. The no-nonsense commercial orientation of ITT has carried over into FCh operations.

One of the initial FCh activities focused on preparation and publication of a directory of Chilean processors and exporters of fishery products. The inventory provided a solid departure for planning FCh fishery related projects, and represented a first attempt at advertising FCh capabilities.

In 1977 James Parkhill, a former sales executive for the Green Giant food corporation, came to FCh to help guide marketing efforts of the subsidiary enterprises. The Third World pace soon proved too slow for this impatient and dynamic executive. The successor was not a hit; the marketing function had to be put on a back burner.

In the early 1980s, Dr. Roberto Echeverría, a Chilean economist with the World Bank in Washington, D.C., was enticed to become the head of FCh's Project Department. There he established a system for project identification, selection, preparation, and evaluation. Marketing implicitly became a component of his terms of reference. As the forestry related projects grew into an ever more important FCh activity, they were elevated to department status.

By 1982 a separate Marketing Department was set up. The job of heading the work was given to Albin Eng, an executive from the Nestlé Corporation, a worldwide master marketer, if indeed anybody warrants that title. The relatively small Marketing Department of FCh is geared up for vigorous promotion of FCh technology transfer services. To build sales strength requires: (a) constant search for commercial opportunities based on new technologies; (b) execution of market studies that supply information vital to development of new projects, or that provide a better understanding of pre-determined markets; (c) designing of new products; (d) determination of brands, packaging, and marketing channels; and (e) systematic dissemination of information about progress of FCh projects.

The Marketing Department has done an outstanding job of selling FCh. Its public relations activities consist of a mix of attractive promotional booklets, carefully prepared copy for advertising of FCh products and services, reprints of articles about FCh and by FCh staff, interviews, seminars, short courses, workshops, open house events, invitationals. A particularly effective tool for marketing FCh has been invitational "work luncheons." They are being held at FCh's executive dining room. Carefully selected groups of potential clients are being invited, including diplomats accredited to the Chilean government. The luncheons help establish two-way communication channels with potential suppliers of technology as well as users of FCh capabilities.

Likewise, frequent press conferences with audio and visual media representatives have been getting the FCh story into newspapers, periodicals, on the radio and television. A partial collection of press clippings, gathered by FCh librarian for the first half of 1985, contains about 150 individual items!

The professional staff follows as well as leads the marketing efforts. They suggest, prepare, and conduct workshops; they write for trade publications; they develop strategies for approaching potential clients. In the course of time most professional staff of FCh have become good at marketing, while some are simply excellent. In short, aggressive implementation of marketing strategies has become an integral tool for FCh survival. Every staff member knows that she or he must pitch in. That is why the marketing function has become widely diffused throughout the entire FCh organization. The Marketing Department primarily serves as catalyst, counsel, and quality controller of the selling work that is largely planned and implement by line professional. This is a refreshing difference from FCh's sister agencies and institutions, which tend to subject marketing to benign neglect.

SOURCES AND TRANSFER OF TECHNOLOGY

Most of the initial FCh staff had U.S. backgrounds. The two general directors who led FCh during its first decade of existence were U.S. citizens. Furthermore, senior management staff tended to have graduate level training in U.S. or Canadian universities. Consequently it is not surprising that the basic technology, adapted for 16 out of the total of 27 projects (Chapter 3), were of U.S. and/or Canadian origin.

Table 7.1   Sources of Technology Used by FCh

Primary Source of Technology	Projects Number	% of Total
1. US/Canada	16	59%
2. Switzerland	3	11
3. US/South Africa	2	7
4. New Zealand	2	7
5. France	1	4
6. Israel	1	4
7. Japan	1	4
8. W. Germany	1	4
Total	27	100%

The primary target markets for the products and services generated by the adapted technology are shown in Table 7.2.

Table 7.2   Target Markets for Technology Introduced by FCh

Primary Target Markets	Projects Number	% of Total
1. Local	15	56%
2. Export	10	37
3. Local/Export	2	7
Total	27	100%

The degree of required adaptation of the selected technologies varied greatly from project to project.

Table 7.3 Adaptations Required in Technologies Selected by FCh

Degree of Required Adaptation of Selected Technologies	Project	
	Number	% of Total
1. Selected technology adapted with minor adaptations	14	52%
2. While selected technology was readily adaptable, implementation of the projects required systematic integration of related production, processing and marketing functions	9	33
3. Substantial changes in selected technologies were required, consisting of extensive location-specific research and field testing of resulting adaptations.	4	15
Total	27	100%

The type of FCh technology transfer work evidently consists of 5 percent inspiration and 95 percent perspiration—hard, systematic, unglamorous follow-up work, which frequently taxes patience and tends to wear down initial enthusiasm.

STATUS OF PROJECTS

As of January 1, 1986—the day the job of Director General of FCh passed into Chilean hands—18 projects, or about two-thirds of the 27 total, were considered "established." The remaining were deemed to be at the apparent stages of maturity shown in Table 7.4.

Table 7.4 Status of Projects, January 1, 1986

Status	Projects	
	Number	% of Total
1. Firmly established	13	48%
2. Established	5	18
3. Questionable outlook for success within the next 5 years	5	18
4. Abandoned and/or Reformulated	2	8
5. Successful pilot operations carried out	1	4
6. Preliminary evaluation made	1	4
Total	27	100%

In short, seven projects, or one-fourth of the total, were considered to have a questionable outlook for success within the next 5 years, and/or have been abandoned altogether. It takes courage and statesmanship to admit that wrong trees have been barked up.

PHYSICAL PLANT

While keeping busy with project work, FCh gradually expanded and improved its real estate. At its 5000 m^2 headquarters complex FCh established impressive physical infrastructure consisting of laboratories, a pilot plant for food processing, a state-of-the-art smoking facility, a superb technical library and information center, functional staff offices, ample conference and meeting rooms, a cafeteria, and an international network for search of technologies.

In March 1975 a strong earthquake, registering 8 on the Richter scale, caused serious damage to the Santiago headquarters of FCh. During the repair some offices had to be temporarily moved, while limited pilot plant work was done on premises of the nearby Technology Institute (INTEC). The reconditioning of the buildings made it possible to introduce improvements, which made the facilities more suitable for FCh purposes.

HUMAN RESOURCES

Even more impressive than development of FCh hardware was the impressive backlog of human resource software, which FCh has been able to mobilize.

As of December 31, 1985, an interdisciplinary team of full-time professionals and technicians totaled 102: three Ph.D.s; ten Masters; five (5) Engineers with overseas studies or post-graduate degrees; 23 locally trained engineers; and 36 professionals specialized in other disciplines. Affiliates of FCh had about 200 full-time employees. In such seasonal programs as quality control, FCh employs 50 to 60 inspectors. As a result, FCh provided work directly and indirectly to at least 355 people.

FCh places great emphasis on development of its personnel. This is accomplished through formal training programs in academic and technical centers in Chile and abroad. In 1985, FCh employees spent 4542 hours in training programs and 1270 hours in supplemental education courses held in technical centers abroad, while FCh brought in 26 foreign experts who rendered more than 500 hours of consultancy work. Crucial hands-on in-service training is being done by establishment of work teams for each project. Within these teams foreign experts participate actively, providing a continuous give-and-take exchange of experiences with their Chilean colleagues.

During its first decade of existence the organization of FCh has gone through several metamorphoses. As of December 31, 1976, FCh had 17 full time employees, 7 of them, or 41 percent of the total, were ex-patriates. There were 4 executive positions: Director General and 3 line directors: (a) Finance and Administration, (b) Food and Nutrition, and (c) Electronics and Telecommunications.

A substantial reorganization took place in December 1985 shortly before the post of Director General of FCh passing into Chilean hands. The resulting adjustment in the functional structure gave origin to departments of Agro-Industry; Electronics and Telecommunications; and Marine Resources. These new departments, along with Forestry, comprise the four main operational areas of FCh; departments of Development, Finance and Administration, and Marketing provided required support.

All seven department heads reported directly to the Director General. The 4 associated enterprises (Salmones Antártica, CULTIMAR, PROCARNE, and BERRIES) reported to the Manager of the Development Department.

During the first decade of FCh's existance its staff participated in virtually hundreds of activities. "The First 10 Years" insert of the 1985 FCh Annual Report highlighted a sample of some 86 happenings: 28 carried out during the initial 5-year period 1976-80 and 58 during the second 5-year period 1981-85 (Appendix 1). At the end of the decade an inventory of about 27 projects, or "cases" in ITT lingo, remained (Chapter 3).

In some cases FCh staff formulated complete projects all the way from product definition to design engineering and industry start-up. In others, FCh staff contribution consisted of formulation of new products and services, improvement of existing products and production processes, making better use of under-utilized local natural resources, as well as training personnel in use and handling of new technologies.

The inventory of projects, reviewed in Chapter 3, shows that FCh staff concentrated activities on sectors based on exploitation of renewable natural resources in agriculture, fisheries, and forestry. These are areas of endeavor in which Chile tends to enjoy comparative advantage over its potential competitors within Latin America and beyond.

SOCIOECONOMIC BENEFITS

During 1985 FCh generated revenues of Ch$ 307 million, which represented about 39 percent of the total operation costs of Ch$ 628 million. In addition to its direct income, which can be readily measured, FCh estimates that it generated about Ch$ 405 million of social benefits, consisting of: (a) increased income among client enterprises, who successfully introduced FCh transferred technologies; and (b) savings of consumers, due to lower prices charged by client enterprises for their new technology-based goods and services. The Ch$ 307 million income and Ch$ 405 million social benefits added up to about Ch$ 712 million. In short, by 1985 FCh "paid its way." That is not bad considering that most similar sister institutions and agencies in Latin America are being permanently subsidized with public funds.

GOING BEYOND CHILE

This summary chapter can be fittingly concluded with a truly pioneering event that occurred in 1985. FCh dipped its toes into work beyond the Chilean borders. Specifically then FCh signed its first international agreement to provide the Foundation for Economic and Social Development of El Salvador (FUSADES), with agroindustry technology. FCh carried out technical and market studies to identify vegetables with export potential as well as varieties which El Salvador can produce as substitutes for imports from other Central American countries.

The pioneering agreement with FUSADES—others have been subsequently negotiated with several countries in Latin America—constitutes a new area of work for FCh. Furthermore, it is a recognition, by the international business community, of FCh's technological capacity and professionalism.

So much then for FCh's track record during its first decade of activities. In the next chapter we draw some conclusions about attributes that clearly distinguish FCh from its 150 odd sister institutions and agencies in Latin America and the Caribbean.

Chilenization of Management

By January 1, 1986 all expatriates—including Mr. Wayne Sandvig, the Director General—returned home. Dr. Anthony Wylie Walbaum was named the Director General. Tony has impecable credentials, holding an undergraduate degree in Agronomy from the Catholic University of Chile, a master degree in Horticulture, and a Ph.D. in Plant Physiology from the University of California (Davis).

During 1970-72 he served as Research Director, Agronomy Faculty, Catholic University of Chile. Between 1972 and 1974 he was Professor of Post-Harvest Physiology, University of Chile, and in 1974-79 General Manager of Fruit Growers Cooperative in the Central Region (FRUGEN).

In 1979 he joined FCh as Director of Food Technology Department. In 1984 he became Deputy Director of FCh and in 1986 full fledged General Director. Dr. Wylie has published widely on matters of his expertise. He speaks Spanish, French, and English, is married to Valerie Moiri, and has 3 children (Caroline, Christian and Susan). FCh appears to "be in good hands", managerially speaking.

ITT, in its advisory role, continues to make available its worldwide experience and expertise. Yet it leaves FCh free to go outside to obtain any special assistance it may need.

Incidentally, on October 13, 1986, Mr. Wayne Sandvig was decorated with the Bernardo O'Higgins order of Knight Commanders, the highest award given by the Government of Chile to a foreign citizen. The citation focused on recognition of his contributions to development of technology in Chile and for his role as a spokesman for the American business community in Chile.

NOTE

1. The award was established in 1956 to recognize foreign nationals who have rendered special services to Chile. There are 4 categories: grand officer, knight commander, officer, and knight. Granted by the president of Chile, on recommendation of the Minister of Foreign Relations, the medal consists of two gold five-point stars, one on top of the other. In the center it carries a medallion with the effigy of Bernardo O'Higgins. It is worn around the neck by a blue and red band. Bernardo O'Higgins, the founding father of Chile, was a soldier who fought for liberation of the country from the Spaniards in 1810.

8

FCh's Relevance as a Precedent
for Sister Organizations

In the last 1976 session of the Board of Directors one
of the Members pointed out that the Foundation could be
considered to have successfully reached its goals if
and when it would become self-financing and operations
be sustainable over time.
By 1985 this goal has been met. We have created
and consolidated an enterprise which has total income
exceeding costs. In addition FCh is generating substantial benefits to the community at large... FCh is a
strong institution which has conquered a major share of
the Chilean technology market. All this has been
achieved with the help of people of exceptional
quality and dedication. I am proud to have been
privileged to direct them. At the same time I am sure
that Dr. Wylie, my successor, will know how to elevate
the institution to even higher levels of achievement.

M. Wayne Sandvig,
Director General of FCh, 1977-85

These excerpts from Mr. Sandvig's eloquent farewell address
to the FCh Board of Directors puts in a nutshell the achievements
of FCh during its first decade in business. The track record is
impressive.

TRACK RECORD

By 1986 the core activities of FCh in food and agriculture,
forestry and fisheries became priorities in the national development agenda. Five commercial enterprises, using FCh adapted

technologies—Berries, La Union, PROCARNE, Caprilac, Salmones Antártica, and Cultimar—were in orbit, and another four companies were approaching the launching pad. The microcomputer work was huge and growing.

The reputation FCh had in the Chilean business community was excellent, and spreading abroad. FCh was attracting top professionals, who considered FCh a prestigious place of employment. Income from service fees was already covering about 40 percent of expenditures. The Chilean Society for Technology Development (SOTEC) served as an effective as well as professional and political support and outreach mechanism for FCh.

The quality certification program became a runaway success, spreading from fruits and vegetables to fish, wine, and wood products. No end was in sight.

Wood has been rediscovered as a suitable construction material. Mobilization of funds needed for expansion of the renewable forestry related resources have been initiated.

FCh staffers have become masterful marketers of FCh goods and services.

More than 200 consultants, who were hired by FCh for specific assignments, constituted a world-wide outreach network of loyal alumni ready and able to help in the search for transferable technologies.

FCh not only assisted in transfer and adaptation of proven technologies, but occasionally took a chance on highly promising but controversial ventures. Irradiation of smoked salmon is a case in point. The gamble seemed fully justified in view of the recent favorable rulings on the matter by the US Food and Drug Administration.

With 20/20 hindsight it is possible to discover some inevitable flies in the ointment. The food canning industry program (CAT) came too early and was put on the back burner. Kiwifruit and jojoba developments have been low-profiled. The initially highly promising production of quality goat cheese (Caprilac) fell by the wayside simply because it proved too difficult to get a reliable supply of raw goat milk. The lupine/wheat rotation did not pass the feasibility test, and was duly abandoned.

FCh management fully acknowledges these flaws and added them to the backlog of "don't" experiences. Fortunately, successes by far outshone these minor blemishes. This made it easy to highlight positive achievements,and reinforce determination of others to emulate, adapt, and hopefully reinforce the best in the FCh precedent.

WHAT MAKES FCh A DESIRABLE PRECEDENT FOR ITS SISTER ORGANIZATIONS?

FCh does not operate in a vacuum. Competition is keen. A survey of food technology resources in Latin America, sponsored in the mid-1970s by USAID, obtained information on 150 sister organizations. Twenty were in Chile: 14 located at universities while 6 were government supported entities. They employed permanent staffs of about 800: 342 professionals, 130 technicians, and 328 in administrative positions. Staff size in these organizations ranged all the way from less than 5 to over 300.

Conspicuous by its absence from the USAID listings is the only potentially substantial FCh competition: Technological Institute of the Chilean Production Development Corporation (INTEC/CORFO). Located a few blocks below FCh in the Parque Profesional, INTEC has been in operation since 1968.

Initially INTEC/CORFO concentrated on implementing R&D project while providing related technical services for food, chemical, metal, mechanical, and electronic industries. During the 1973-83 period INTEC expanded its activities into such promising new areas as exports of agricultural products, upgrading of coal mining, environmental control, materials handling, and information systems.

While FCh has apparently been successful, many of the 30 other Chilean sister institutions linger on, understaffed, underfinanced, and unappreciated by their clienteles. Why? What combination or mix of distinctive and identifiable features have contributed to FCh's achievements? What were the catalysts that triggered FCh activities, fueled them, kept them on track, accelerated them, or slowed them down or stopped them altogether, if need be? FCh's genesis, combined with an analysis of the 27 projects (Chapter 3), revealed at least 7 substantive features that might help explain what makes FCh so different: circumstances of birth, management by objectives, skilled technology marketing, target oriented flexibility, prudent risk taking, access to a wide technology network, and capability to prepare packages of bankable projects. In this case study, FCh's strengths were intentionally highlighted in a way that hopefully would encourage sister institutions to consider selective emulation and/or adaptation of this desirable precedent. In short, I wanted to reduce the "Yes, but we are different" reaction, which is often used as a rationale for avoiding a precedent, no matter how desirable or relevant. Indeed, I maintain that not even the circumstances surrounding the conception and birth of FCh were that unique. Let me explain.

UNIQUE ORIGIN

In a November 1984 interview with editors of the Chilean Management Review (Gestion), M. Wayne Sandvig, then Director General of FCh, categorically stated: "Fundación Chile is unique not only in Chile, but in the entire world. This is because it is a joint venture between a goverment and a transnational company aimed at systematically incorporating new technologies into the country's economy."

Uniqueness alone does not necessarily make for desirable precedent. What is genuinely unique about the relationship between the government of Chile and ITT are the circumstances under which FCh was created. A politically sticky dispute over compensating ITT for its nationalised properties had been going on for years. No satisfactory resolution of the problem seemed to be in sight until 1974, when Raúl Saez, Chilean Minister of Economic Coordination, got the brilliant idea of creating FCh. That proved a most unique and imaginative way to "turn a sow's ear into a silk purse." Those circumstances cannot be duplicated. Yet the creative and positive approach to an apparently intractable problem is the lesson of unconventional wisdom that should be derived from this Chilean precedent.

HUMAN RESOURCES

The high productivity of the relatively small professional staff is an outcome of FCh's ability to attract and keep good people. FCh soon became a desirable and prestigious employer, able to attract some of the best local professionals as well as bring back from abroad expatriate Chileans (a brain gain).

FCh "stretches" its staff capabilities with frequent short-term contracts with experts from around the world. In 1984 alone some 33 consultants were contracted for their expertise in the following areas of technological endeavors: meat, 1; forestry and wood processing, 9; cheese, 8; wine and viticulture, 4; asparagus, 3; food irradiation, 2; fisheries, 2; and one each for castor beans, fruit, ginseng, meat, and vegetables.

For its permanent staff FCh has job descriptions that specifically and precisely define their task; these documents are reviewed annually. Likewise, explicit definition of objectives and carefully drafted terms of reference facilitate identification and selection of consultants in specific fields of needed technologies. The outreach toward a worldwide pool of talent is made possible by FCh's unique international linkages, one of its major comparative advantages.

MANAGEMENT BY OBJECTIVES

In the late 1970s the long-established ITT R&D managment by objective (MBO) system was adapted to Chilean requirements for systematic management of technology transfer (MOTT). That disciplined approach to managment requires teamwork, a matter that prior to the 1970s clearly militated against technical research as traditionally practiced in most Chilean public institutions. Even though academic research work was being partially or fully financed by public funds, professionals tended to view their research topics and methodology as exclusive property. Researchers were more eager to duplicate investigations recently published abroad, than to solve practical Chilean problems. Research was aimed at sacred intellectual themes. To channel science and technology toward solutions of entrepreneurial problems was considered heresy. Scarce resources tended to be aimlessly allocated to a wide variety of irrelevant research topics. The style of research was one-man ventures, mitigating against teamwork. Consequently the government's research investments were mainly acts of good faith, rather than goal-directed efforts. This climate pervaded individual technical universities, agricultural colleges, and experimental stations as well as industrial research institutes. Few if any individual enterprises did any applied research. Mutual suspicion pervaded the relations between Chilean industry and academia. There was little dialogue that would help the business community articulate relevant problems applied researchers could and should help resolve. The rise and fall of the CAT project, aimed at improving quality and productivity in the Chilean food canning industry, was due to lack of grassroots involvement with defining the problems of the industry, or grassroots participation in planning and execution of the program—that is, poor marketing.

Within a decade teamwork became routine within FCh. As effective dialogues and cooperative ventures were being carried out by FCh with Chilean academicians, industrialists, bankers, businessmen, and government officials, the team approach is rapidly spreading beyond the confines of FCh.

TECHNOLOGY MARKETING

Considering that for its entire length Chile borders on the Pacific Ocean, it was only natural that a survey of Chile's marine resources and their utilization was among the early FCh activities. This work was summarized in directories of: (a) commercial fishery firms and their key personnel; (b) equipment and facilities used for harvesting, processing, and storage of fish; and (c) methods of fish marketing. This data base helped shape programs for development and expansion of Chile's fish exports as well as marketing fresh fish within Chile. Similar export promotion work was done for selected fruits, vegetables, wine, berries, and forestry products.

The fullfledged Marketing Department, which was established in 1983, has served as a catalyst to inspire and guide FCh staff in developing and implementing their own strategies for marketing FCh products and services. And of course the Marketing Department was doing its share of marketing research for in-house uses as well as for third party clients. All that was pretty routine. Far less routine were the imaginative ways the Marketing Department acted as master strategist and tactician in selling FCh to crucial potential clients as well as to professional and political allies. The most valuable advertising agent for FCh was the word-of-mouth from satisfied clients of FCh or partners in joint venture enterprises. Academicians were increasingly being drawn in through research and training subcontracts. A well managed 450 member Society for Technology Development (SOTEC) grew into an effective outreach tool spreading technology consciousness within the country, while strenghtening FCh's network of local experts that could be mobilized for work on specific FCh projects.

FCh "invitationals" were among the least publicized, but most highly targeted public relations devices used by FCh. These events usually consisted of work lunches and tours of head-quarters and/or field trips to FCh enterprises. Invitations went to local and foreign decision makers related to technology transfer: entrepreneurs, academicians, politicians, military personnel, journalists, diplomats, professionals, members of civic associations, and other community leaders. All these target individuals needed to know what the FCh was doing, and were to establish personal acquaintances with key FCh staff. Lunches were held in FCh's attractive executive dining room. Menus included products from FCh subsidiaries or clients. Invitations to these events have become a status symbol of sorts. During 1984 some 37 such events were held, attended by about 335 individuals.

The fact that the marketing approach to technology transfer has been made to work in FCh sets a highly desirable precedent for all of Chile and beyond. This is because, as incredible as it sounds, among the about 150 food technology related insti-tutions in Latin America only one seems to be systematically and explicitly moving toward a market orientation: the National Brazilian Research Center for Agroindustrial Food Technology (Centro Nacional de Pesquisa en Tecnología Agroindustrial de Alimentos, CTAA).

The affinity between CTAA and FCh lends itself to active cross-pollination or "institutional twining" as the World Bank chooses to call the phenomenon (see Chapter 9 on Roads Ahead).

TARGET-ORIENTED FLEXIBILITY

Within selected action areas FCh is ready to tackle almost any barrier that hinders effective transfer of technologies aimed at improvement of the production and marketing processes. As the reputation of FCh has grown there have been many temptations for deviating from the initial core of 4 key activities: food and agriculture; fisheries; forestry; microcomputers and telecom-munications. Fortunately these temptations have so far been successfully resisted. FCh does not intend to be weakened by pretending to become a "jack of all trades" and presumably master of none.

The self-imposed limitations do not apply geographically. Within Chile FCh has ongoing activities in at least ten out of the twelve administrative regions of Chile. This widespread local presence makes FCh increasingly visible and known throughout Chile. Furthermore, when in 1985 an unexpected opportunity arose for reaching beyond Chile, FCh management promptly decided to undertake a pioneering export of its skills and experience to El Salvador.

Equally important is the undoctrinaire flexibility of FCh in implementing ongoing projects. The systematic and continuous monitoring of the work progress provides specific feedback that is quickly being translated into remedial actions. FCh has thus abandoned projects that proved ineffective, backtracked on others, and put still other cases on a temporary hold, when it appeared that the FCh inputs were premature. Likewise, there seems nothing sacred about FCh's organization chart: responsibilities are being redistributed, department and work groups created, merged, eliminated, increased or decreased, all in accordance with ever-changing requirements of the institution. Terms of reference and job descriptions for key professionals are periodically being reviewed and updated.

In one respect FCh is stubbornly inflexible: maintenance of a high quality of work.

ACCESS TO TECHNOLOGY NETWORKS

Jean-Baptiste Voltaire's dictum to the effect that "to imitate is the highest form of originality" can only be effectively practiced by those who know where to find technologies worth imitating.

FCh's association with ITT has provided effective access to a worldwide network of consultants able to transfer technology to Chile just as effectively as bees carry pollen from flower to flower. Among the key mechanisms that guide the process of orderly technology transfer is the 5-year Technical Assistance Contract (TAC) signed on November 14, 1985, by FCh and the International Standard Electric Corporation (ISEC), a subsidiary of ITT. The new TAC superseded the initial 10-year Management and Technical Assistance Contract (MATAC), signed on August 27, 1976.

Within this framework, TAC would continue the reciprocal applied technological research and technology transfer programs in the four priority fields: agro-industry, forestry, marine resources, and electronics.

The contract commits FCh to: (a) reimburse ITT for all direct costs incurred in providing technical information, including cost of copying documents and reasonable travel, hotel, and meal costs of any personnel required to provide technical assistance at FCh premises; (b) keep the data confidential until different use is authorized in writing; and (c) promptly inform ITT of any invention, discovery, or improvement that may result from FCh's research programs conducted under the framework of TAC.

Likewise, if an FCh technology were to be based on research, using substantial and significant amounts of ITT technical information, ITT would be free to use such invention, discovery, or improvement, pay FCh all direct expenditure it may have incurred in transferring such information, including reasonable travel costs, subsistence expenditure, and remuneration of FCh personnel providing technical assistance to ITT.

In an attempt to make fruits of the FCh/ITT partnership widely available, ITT committed itself to consider issuing non-exclusive rights to FCh licenses for Chile as well as for other Latin American countries and beyond.

In addition to TAC, the more than 200 international consultant, who have previously been contracted by FCh constitute a valuable alumni association that is being effectively used in the search for needed new technologies.

Within Chile itself FCh has ready access to the more than 450 members of the Chilean Society for Technology Development (SOTEC), which FCh helped create. SOTEC provides an effective communication channel with technology related professionals in local government agencies, sister research institutions, academic personnel, business and trade organizations, as well as technical personnel in private and public enterprises.

To establish and nurture such networks does not cost much money. But it does require imagination, patience, and the conviction that open cross-pollination of technologies tends to generate far more important benefits than pathological secretiveness that has characterized many professionals working in FCh sister organizations.

PRIVATE INITIATIVE

Since its inception FCh has put emphasise on development of private initiative. In the words of Gustavo Chiang, Director of FCh's Forestry Department:

> Our aim is to acquaint ourselves with the problems faced by Chilean private enterprise which hinder its development, and work together with the industry in finding practical solutions. This might require an investment by Fundación Chile at one stage or other. But it should be clearly understood that we intend to take no steps that might interfere with private interests or compete against them. Quite the contrary: we wish to support private enterprise by helping to locate, within the country or abroad, technologies... that may be economically applicable in Chile.

The contemplated sale of affiliated FCh enterprises to local investors fits in well with the vigorously enforced national privatization policy, which aims to systematically reduce participation of the government in the economy (see Chapter 10, Epilogue).

One of the desirable by-products of the privatization drive was emergence of a genuine internal capital market. However, by the mid-1980s purchase by social security funds of common stock in companies affiliated with FCh appeared on the risky side. Other sources of funding had to be explored. Consequently in 1984 FCh undertook a feasibility study for establishment of venture capital companies.

In 1986 CORFO invited FCh, the World Bank's International Finance Corporation (IFC), and a group of Chilean businessmen to form a Venture Capital Fund aimed at stimulating domestic private enterprise. Initial funding is to be between at CH$1.6-2 billion. The fund is to be managed by an administration company responsible for selecting attractive projects with an adequate internal rate of return (IRR).

In short, FCh has become involved in adaptation of technology aimed at facilitating privatization of government-owned enterprises. Since the early 1980s the privatization fashion has been rapidly spreading from Europe and North America to Latin America and the Caribbean. Most countries in the region have established privatization policies and are setting up implementation mechanisms. The lack of effective capital markets has been one of the most serious constraints to rapid privatization. Under these circumstances FCh skills in this field of endeavor should find ready markets within Chile as well as abroad.

GENERATOR OF BANKABLE PROJECTS

No matter how brilliantly conceived is an idea for technology transfer or a technology based enterprise, eventually somebody will have to foot the bill for implementation of the project. Rigorous project preparation and evaluation skills are insufficiently diffused in Chile and beyond. Consequently, since the mid-1975 countries in Latin America and the Caribbean have increasingly suffered what has come to be called a drought of potentially bankable projects. FCh enjoys a comparative advantage over many of its potential competitors in that it has a growing number of professionals seasoned in the art of hardnosed economic and financial evaluation of projects at different stages of their life cycle, all the way from conception of ideas, through preliminary studies, pre-feasibility and feasibility work to ability to search for joint venture entrepreneurs inclined to risk capital in promising innovative activities. Indeed Dr. Robert Echeverría, a former economist with the International Bank for Reconstruction and Development (World Bank) in Washington, D.C., helped FCh steer in the direction of generating "bankable projects," including packaging of financing needed to get the ventures into orbit.

The ability of FCh staff to prepare documentation needed for making projects bankable is a rare skill in developing countries. Rather than suffer a "draught of bankable projects," FCh has a substantial backlog of projects, with corresponding financing packages, waiting for investors. Furthermore FCh is increasingly putting its project preparation skills at the disposal of third parties.

Other observers and students of FCh are undoubteadly able to identify other skills that warrant emulation or adaptation within as well as beyond the borders of Chile.

In the next chapter I explore some alternative roads ahead for FCh on its journey during the second decade of its activities.

NOTE

1. CTAA is one of the 24 centers of the Brazilian Public Corporation for Agricultural Research (Empresa Brasileira de Pesquisa Agropecuaria, EMBRAPA). EMBRAPA was established in 1972 for the purpose of increasing quantity and quality of scientific knowledge relevant to agricultural development. EMBRAPA guides applied agricultural research in accordance with concrete needs expressed in government policies as well as concerns articulated by farmers, extension agents, and industry. In short, EMBRAPA aims to develop technologies designed to increase productivity of agriculture, expand domestic food and fiber supply, increase farm income, and improve quality of life in rural areas.

Early in 1985 CTAA hired Professor A.J. Alton from the University of Rhode Island to prepare a plan for marketing technology transfer to the Brazilian agroindustry. The terms of reference for Prof. Alton explicitly called for an individual familiar with marketing of technology and information to: (a) evaluate technological needs, reflecting marketing opportunities for the Brazilian food industry; and (b) help establish the institutional capacity of the EMBRAPA research network to produce and transfer results of its research to the food industry.

See: Aaron J. Alton (Department of Marketing, College of Business Administration, University of Rhode Island, Kingston, RI 02881; telephone (401)792-2083), Transfer of Technology—Marketing in Disguise: The Case of CTAA in Brazil. Paper presented to the 10th Annual Macromarketing Seminar, Airlie Conference Center (Virginia), August 15-18, 1985.

9

Roads Ahead

Chile's long-range economic welfare lies more in the direction of sophisticated science and technology-based industries rather than merely in exports of raw materials from mining, forestry, and agriculture. To switch gears toward the required technology orientation takes time and vision.

Unfortunately in the last ten years, FCh has not even come to first base on that score. And that is not for lack of trying. Some of our friends believe that we are simply too far ahead of time. I heartily disagree. If anything, Chile lags behind in comparison to similarly endowed Third World countries that have come a long way into restructuring their economies into high-tech directions: Brazil, Israel, Singapore, South Korea, Taiwan... These are just a few examples of the so called newly-industrialized countries. I hope that, within the next decade or so, FCh will act as one of the strong catalysts that would help push Chile into the NIC group.

Author's interview with M. Wayne Sandvig, Director General of FCh, 1977-85

After its first ten years of operations FCh has proven its capability to offer a wide range of technological services to national industry: food formulation, quality control, and certification for exports of fruit and vegetables, process automation with use of microcomputers, development of rural telephone networks, chemical and microbiological laboratory services, operation of a pilot and demonstration plant for food processing, intensive use of forestry resources, pragmatic methodology for economic and commercial project evaluations,

advisory services, and technological assistance for export
development. In short, FCh's well-balanced staff of scientists,
engineers, economists, and market specialists has the capacity to
enhance technological development of selected industries for the
benefit of all Chileans.

The services rendered by FCh are apparently appreciated by
the Chilean business community. In 1985, FCh revenues already
covered 39 percent of operating costs.

On January 1, 1986, when the post of Director General passed
into Chilean hands, executive directors and staff of FCh outlined
the activities aimed at generating substantial economic and
social benefits during the second decade of FCh's existence. In
the following sections I will review the roads ahead, as visual-
ized by these FCh insiders, and in addition add my own unsolic-
ited visions.

VIEWS OF INSIDERS

On December 17, 1985, M. Wayne Sandvig, Director General of
FCh during 1977-85, gave his farewell address to the FCh Board of
Directors. He outlined the roads ahead (el camino por recorrer)
by offering some advice aimed at assuring continuity of FCh while
increasing its socioeconomic impact. By and large Mr. Sandvig
foresaw more of the same. He highlighted technology as a
fundamental ingredient in development of Chile's export-oriented
market economy. Consequently, he recommended that FCh activities
should always be linked to helping Chile achieve that objective.
This would call for FCh to maintain a dynamic, agile, and
flexible organization capable of using its established expertise
within the productive sectors of the economy.

FCh's most valuable asset will always be its technical
staff, at headquarters and in the field. Investment in continued
in-service training of this human resource therefore needs to be
continued, intensified, and complemented with a substantial dose
of outside seminars, travel, and consultancies. Likewise, the
crucial catalytic actions of FCh need to be amplified.

In order to improve the track record of FCh during 1986-90
Mr. Sandvig suggested a minimum of two specific goals.

First, provide technical assistance services, culminating
with FCh clients establishing and investing in at least one
additional associated enterprise per year. This would require a
combination of: (a) selective use of foreign experts of proven
professional expertise, and (b) assertive marketing, to existing
and potential clients, of FCh's technical skills. Among the
tools that this type of marketing effort must utilize are short
courses, workshops, seminars, publications, open houses, visits
to FCh headquarters and enterprises, and public relations with
mass audio and visual media.

Second, establish within and outside FCh, policies and
incentive systems favoring technological innovation based on

private initiative, including mechanisms for provision of
risk capital on more favorable conditions than traditional
projects can obtain in existing capital markets.

Mr. Sandvig then quantified the benefits FCh might conceiv-
ably be able to produce during its second decade of operation.
His projections were based on three conservative assumptions:
(a) FCh could generate social benefits at an annual rate of some
Ch$ 665 million, which was experienced during the 4-year period
1982-85; (b) FCh should cease operation by 1995 without leaving
behind any residual value; and (c) the total accumulated net
benefits should be discounted at 12 percent annually. The
resulting total benefits added up to some Ch$ 3.5 billion,
indicating a 24 percent return on investment in FCh projects, and
48 percent return on investments made in FCh by the government of
Chile. Should FCh be able to continue operations successfully
beyond 1995, with results at least as favorable as the assumed
targets, the return on the investment would be even larger.

The ink has hardly dry on Mr. Sandvig's message to the FCh
Board of Directors when the roads ahead started to be widened.
In early 1986 FCh institutionalized its outreach beyond the
borders of Chile. Dr. Robert Echeverría made an exploratory trip
around South and North America. He visited multilateral and
bilateral development agencies in Washington, D.C. (USAID, World
Bank, IDB, OAS) to ascertain the chances for FCh services to
participate successfully in international competitive biddings.
The recommendation to the Board of Directors was affirmative:
FCh appears capable of systematically expanding its activities
beyond the borders of Chile. The proposal was duly accepted and
Dr. Echeverría was named Director of FCh International. This
budding export-orientated offshoot deviates from the original
intent of having FCh serve as bridge for bringing technology to
Chile.

AN OUTSIDER's PERSPECTIVE

My suggestions aim at broadening and deepening FCh's
catalytic work, which I consider among the most rewarding
functions that can and should be performed by such an organization. None of my suggestions are wild shots into the dark.
Rather, they are based on FCh's achievements during its first ten
years of experience and/or proposals that have been previously
made, but, for various reasons, not acted upon.

The prime target within Chile for the four enumerated areas
of promising FCh catalysis are: packaging of bankable technology-based projects; technology transfer consortia; incubators of
technology-based enterprises; and fostering of technology-
friendly mentality among Chilean entrepreneurs.

Three proposed additional activities would primarily concern
potential customers of FCh services, outside the borders of

Chile: inscription in international rosters of consulting firms; restoration of links to California; and internationalization of the Chilean Society for Technology Development (SOTEC).

The reader can readily add many more suggestions for alternative roads ahead as well as come up with variations on existing themes.

WITHIN CHILE

Packaging Bankable Projects

Transferred technology cannot benefit anybody until put to use for personal or commercial purposes. Implementation costs money. Full-fledged feasibility studies guide potential investors in their decision making. Chile has limited experience in preparing bankable technology transfer projects. But FCh has an established track record that can form a basis for a substantial expansion of activities related to packaging of bankable projects and serving as intermediary in the search for potential providers of technology and/or investment funds.

Cooperation with CORFO's small enterprise development programs, in which FCh already actively participates, is one outlet for FCh's catalytic skills.

New vistas have also been opened with the Inter-American Development Bank's U.S.$32.5 million program for research and transfer of agricultural technology. The four major purposes of the loan are to:

(a) increase agricultural production by helping to generate and adapt technology for priority products destined for local consumption as well as export;

(b) strengthen research aimed at improving location-specific use of such natural resources as soil and water;

(c) improve the system and mechanisms for transfer of technology, so as to put obtained research results at the service of all agriculture; and

(d) develop production systems that will improve the efficiency of the sector by reducing cost and using available resources more effectively.

The program would address adaptive research needs in crops as well as in livestock.

Among the rewarding opportunities for packaging bankable projects is the crucial forestry sector, in which FCh has substantial proven capability. In 1986, largely as a result of FCh's catalytic activities, a Forestry Investment Committee (FIC) was formed with a mandate to "look for ways to maximize Chile's comparative advantage... by anticipating expansion and making certain that essential infrastructure—such as ports, roads, transport, facilities, as well as financing—be available when needed." Based on the large backlog of data gathered on the forestry sector, FCh should become one of the major packagers of potentially bankable projects in that sector.

Technology-Transfer Consortia

Even with the aid of outside consultants, FCh can only master a limited number of activities. If FCh were to create R&D consortia with some of its Chilean sister institutions the scope of these activities could be broadened substantially. Benefits from economies of scale in technology adaptation and transfer could be enhanced by concentration of consortia members within a specific location. The Research Triangle (Raleigh, Chapel Hill, and Durham) in North Carolina is a typical example of geographic concentration of R&D institutions. Another case in point is the Tsukuba Science City in Japan, which in 1986 consisted of a cluster of over 50 government and private research institutes and two universities.

A point of departure for creation of such a potential R&D triangle already exists in the Parque Institucional in Santiago. There, within a few blocks of each other, three institutions are already headquartered: FCh, the National Research Institute for Metals and Metallurgy (INMM), and the Chilean Technology Institute (INTEC), a subsidiary of CORFO. Precedent for this sort of active cooperation was created in 1985/86 when the FCh pilot plant for food technology, damaged by the March 1985 earthquake, temporarily operated on the premises of the nearby INTEC.

Incubator of Technology Based Enterprises

So far FCh has spread its activities, through most of Chile's 12 administrative regions (Appendix 3). That is as it should be, because initially FCh needed to establish its presence throughout the country.

In years to come economies of scale, generated by concentration of activities in common or close-by facilities, might add to FCh's capacity for promoting small science- and technology-based enterprises. The strategy is known as "industrial incubation."[1]

Industrial parks are already operating in several Chilean localities. However, none is specializing in technology-based activities. Physical proximity tends to generate symbiotic effects for risk capital ventures. FCh might be able to entice potential entrepreneurs to invest in fledgling enterprises if they were to be located in an area where they could benefit from management counsel aimed at increasing their survival rate during the critical initial 2-3 years. The experience with its five associated companies qualifies FCh to catalyze creation of such an enterprise incubator. Indeed, some economies of scale for participants in the PROCARNE meat packing enterprise have already been created in that several of the livestock farmers are also stockholders in the Berries La Union. This "fusion", although unplanned, should be institutionalized wherever geographic and technical circumstances prove favorable.

Fostering Technology Friendly Mentality

In one of the most fascinating sessions that I had with Mr. Sandvig, he revealed some "great visions" for Chile. Mr. Sandvig was thus convinced that with its superb human resources, Chile should and could become a newly-industrialized country, thereby joining the club with such countries as Brazil, Israel, Singapore, South Korea, and Taiwan. According to Mr. Sandvig, the major barrier to this achievement was in the hearts and minds of the Chilean entrepreneurs. He repeatedly harped on the need to develop an innovation-friendly mentality (IFM) among leaders in the Chilean business community.

I fully agree and suggest that SOTEC add an IFM public relations campaign to its agenda of priority work, to be tackled by the late 1980s. FCh's Marketing Department can readily develop effective strategies aimed at achieving this crucial change toward innovation among Chilean entrepreneurs.

GOING BEYOND CHILE

During its first ten years of existence, FCh has been primarily an importer of technology. By 1985, when awarded a contract for a series of studies of the fruit and vegetable sector of El Salvador, FCh successfully tried activities beyond the borders of Chile. This means that FCh turned into exporter of technology (i.e., helped to "enhance commerce and the flow of technology"). Here are some ideas for systematic broadening of FCh outreach beyond the borders of Chile.

International Rosters of Consultancy Firms (DACON)

As a result of the debt crisis of the mid-1980s an increasing proportion of private and public development projects are being funded by multinational technical and financial assistance agencies such as the United Nations Development Program (UNDP), the World Bank, and regional development banks in Latin America, the Caribbean, Asia, and Africa. In order to become eligible for work with those institutions and their clients, FCh must deposit DACON Forms 1600, which are designed for computerized registration of consulting firms with the Inter-American Development Bank, and other international financial institutions.[2]

Considering its successful USAID-sponsored 1985 consultancy in El Salvador, FCh should also register with USAID in the Bureau for Private Enterprise (PRE) roster of resources able to assist in implementing private enterprise initiatives in developing countries. The PRE roster specifically strives to identify skills in: (a) business management and entrepreneurship development; (b) capital markets and financial institutions; (c) rationalization and divestiture of parastatal enterprises, (d) agricultural and agribusiness development; (e) promotion of exports and investment; (f) transfer and marketing of appropriate technology; and (g) development of industry, manufacturing, and nontraditional exports. (The PRE roster has been compiled by Coopers & Lybrand, 1800 M Street, Washington, D.C. 20036).

Many FCh competitors who are already in the international rosters of consultancy firms have mastered the art of preparing studies and reports on pre-feasibility and feasibility projects. Few consultants know how to make projects marketable or bankable, that is, presented in form and with substance adapted to customary requirements of such international technical and financial agencies as USAID, FAO, regional development banks, the World Bank, and potential joint venture investors. In establishing its associated enterprises and helping to prepare the recently consummated joint ventures in the Chilean forestry sector, FCh has proven its capabilities. This comparative advantage of FCh has to be highlighted in the DACON Form 1600.

This is because since the early 1980s, private and public bilateral and multilateral development agencies have found it more and more difficult to place available funds in socioeconomically attractive projects. The buzz words of "project drought" were incorporated into the dictionary of the development profession.

Agriculture, forestry, and fisheries—sectors afforded high priority in practically all developing countries—are particularly hard hit by the drought of bankable projects.

A combination of factors have brought about the project drought: (a) harder loan conditions, with higher interest rates and shorter grace and amortization periods; (b) high indebtedness, making LDC finance ministers reluctant to provide guarantees for additional loans; (c) difficulties in mastering local counterpart funds, which are needed to match or complement funds received from international lenders; (d) most standard farm commodities (wheat, rice, corn, soybeans, dairy products, cotton) are emkpuomg worldwide surpluses, making it difficult to justify projects aimed at increasing production of these traditional products; (e) joint ventures, among foreign equity investors and providers of technology for nontraditional export products (winter vegetables, flowers, ornamentals, tropical fruits and roots aquaculture, and other such exotics) and their local counterpart entrepreneurs are difficult to put together because the required technology is not locally known; and (f) the recent emphasis on privatization, including private/public joint

ventures, calls for a spreading of risk over several participants, who might have conflicting objectives making technically
and financially viable project preparation an ever more
challenging task.

LINKS TO CALIFORNIA

In 1986 a five-year Technical Cooperation Agreement was
negotiated between FCh and ITT. The contract gives FCh continuous access to ITT technologies during 1986-90. This type of
arrangement has worked well during the first ten years of FCh's
activities, and it is likely to work satisfactorily during the
second half of the 1990s.
However, it would appear advisable for FCh to diversify its
sources of technology. A first step in this direction could be
restoration of links to California and the Stanford Research
Institute. This is because during the 1960s a unique
USAID-funded Chile-California program created strong linkages
between participating universities and research organizations in
both countries. Stanford Research Institute (now called SRI
International) in Menlo park was coordinator of that huge
program, providing most of the U.S. staff.
By the mid 1980s there was no systematic institutional
memory left for that intimate relationship, established two
decades earlier. The USAID support was phased out and
professionals involved on both sides have long been scattered
around the world. SRI managers, who spearheaded this attempt at
mass transfer of technologies, are retired or departed for
greener pastures.
FCh executives have visited Menlo Park in search of clues
about "care and feeding" of a private self-financing institute
aimed at catalysing technology development and transfer. SRI
executives treated the FCh visitors just like they would treat
any other potential client, to be charged the customary fees.
The time has come to re-establish a special twinning
relationship between those two institutions, especially now when
FCh is actively broadening its horizons. SRI International has
much relevant experience to share. After all, the original SRI
was created shortly after World War II, under the wings of
Stanford University, one of the leading U.S. institutions of
higher learning. The University-pioneered adjacent Stanford
Industrial Park (SIP) flourished and its influence expanded to
the flatlands for miles south of the university, an area nowadays
known as "Silicon Valley." Elements of this enterprise incubator
experience can undoubtedly be adapted by FCh to Chilean
circumstances.
Dr. Anthony Wylie, since January 1, 1986, the Chilean
General Manager of FCh, obtained his Masters and Doctorate from
the University of California (Davis). That is a good connection

on which to rebuild linkages to California in general and to SRI International in particular.

U.S.-CHILE BINATIONAL INDUSTRIAL RESEARCH AND DEVELOPMENT FOUNDATION (BIRD)

FCh does not need to confine the "twinning" relationships to ITT and SRI International. There are many more alternative forms of outreach that should be explored. The U.S. Binational Research and Development Foundations (BIRDs) are a case in point. The first BIRD was set up in 1977 "to stimulate, promote, and support mutually profitable cooperation between private sectors of the United States and Israeli technology industries." To carry out its purpose BIRD used income from a $110 million endowment provided equally by the governments of Israel and the United STates. The U.S. and Israeli participants in BIRD seek to develop and commercialize innovative but non-defensive technical products or processes. BIRD's return on its investment comes from royalties on sales resulting from funded projects. BIRD picks up the tab for unsuccessful projects. By 1985 BIRD had funded over 125 projects; royalties for the year totalled $0.9 million and income from endowments added $5.6 million. In 1986, based on the successful Israel-U.S. BIRD, an Indian Foundation was set up.

A similar U.S.-Israeli Binational Agricultural Research and Development Foundation (BARD) has also been in operation since the mid-1970s.

FCh might want to take the initiative in exploring the feasibility of establishing new symbiotic relations via little Chilean BIRD and/or BARD foundations aimed at multiplying effectiveness of research and development works and accelerating utilization of resulting location-specific technologies.

INTERNATIONALIZE MARKETING OF FCh

The FCh Marketing Department is well equipped to develop goals and components for a long-range campaign aimed at selling FCh services abroad.

Let me suggest just a couple of potential marketing strategies for blowing FCh's professional horns loudly, frequently, and widely. One strategy would aim at selling FCh services to clients abroad, such as was the case with the feasibility study for fruits and vegetable development in El Salvador. The other strategy would aim at completely different and most unconventional markets: twinning with sister institutions and agencies in Latin America and beyond. Both strategies are based on established FCh practices within Chile; they merely need to be externalized. Let me explain.

Blowing the FCh Horn

    Demand for FCh services—a type of supply economics—needs
to be created by intensifying and systematically pursuing
existing international outreach activities (see section 4.4.5).
Much of the increased international visibility of FCh can be
obtained with practically no additional cost. The FCh Marketing
Department can: (a) systematically encourage former consultants
to write up their assignment and publish articles in professional
and trade journals, giving due credit to FCh and/or present them
at professional meetings; (b) place case stories about FCh
enterprises, goods, services and people in international media
such as Agribusiness Worldwide, International Development, South,
Third World Quarterly, International Projects, and so on;
(c) send copies of DACON Forms 1600 to commercial attachés at
Chilean embassies; (d) make copies of English and Spanish
versions of this book available to potential clients abroad, and
to foreign diplomats and business representatives in Chile;
(e) enable importers of products with FCh Quality Certificates to
tell the FCh story, and so on. The sky is the limit as to what
can be done in spreading the FCh story abroad.

Internationalize SOTEC

    All the above enumerated target markets have been of the
conventional type. Now let me suggest a most unconventional
market for FCh services: sister technology research and transfer
institutions and agencies throughout the world.
    Since the early 1960s Latin America and the Caribbean have
experienced a proliferation of new institutions doing
technology-related research and development work. In a 1977
USAID-sponsored study information was gathered on about 147
organizations dealing with food technology in 15 countries (see
Table 9.1).
    If the more than 30 institutes in Mexico and Venezuela were
to be added, the total would likely exceed 180 such places for
food and agriculture alone.
    The institutes vary in size from as few as 2-3 employees all
the way up to hundreds of professionals. They perform a variety
of tasks: teach, train, do research, run pilot plants. Some are
barely surviving; others are thriving. Some do location-specific
assignments, most are national in scope. At least half a dozen
(ICAITI, INCAP, CIMMYT, CIAT, CIP, CFNI, etc.) have multinational
sponsorship. Some are private; most are public.
    There are at least 100-150 more technology-related insti-
tutes engaged in activities other than food and agriculture.
This means that FCh has close to 300 sister institutions with
whom to cross-pollinate ideas, experiences, management styles,
marketing, and outreach strategies— all activities for which no

systematic channels of communications have as yet been established. All that is so far available are accidental international meetings, forums, conferences, workshops, or symposia,
which aim to further interchange of ideas among selected staffs
of these institutions.

Table 9.1 Food Technology Related Institutions in Latin America
and the Caribbean, 1977

Country	Number	Country	Number
1. Argentina	21	9. El Salvador	2
2. Brazil	42	10. Guatemala	4
3. Bolivia	6	11. Jamaica	2
4. Chile	20	12. Nicaragua	1
5. Colombia	15	13. Paraguay	3
6. Costa Rica	1	14. Peru	11
7. Dominican Republic	4	15. Uruguay	9
8. Ecuador	6		
		Total	147

Source: B.F. Buchanan and G. Stewart, Contribution to Food Technology Resources in Selected Latin American Countries.
Washington, D.C.: USAID, May 1977.

Executives, who want to learn about how to run their
technology transfer institutions effectively, have few places to
go for advice. Leading U.S. institutions—Battelle, A.D. Little,
SRI International, Research Triangle, to name a few—cannot
afford the time and staff to freely give away their know-how.
There is no regional professional association that serves that
purpose. Consequently, the diffusion of management skills
related to technology transfer largely remains a hit-and-miss
proposition, a costly learning-by-doing process. Furthermore,
marketing remains a Cinderella function in most of these institutions, causing many of their financial problems.
The time has come for exploring ways in which skills and
experience in management of technology flows could be pooled,
analyzed, interpreted, and made more readily exchangeable among
institutions and agencies in the region. As an initial step,
SOTEC might be able to organize a regional workshop for 10–15
selected key staff members from a limited number of selected
institutions. The purpose of such a gathering would be to define
management challenges faced by public and private technology
transfer institutions in the Western Hemisphere, and to outline
an action program focused on diffusion of effective management
systems among technology transfer institutes.

Multinational and bilateral agencies have a vital interest in furthering effective transfer of technology, as well as a growing backlog of research work already accomplished by them or under their auspices. IDB and/or the World Bank might consider actively co-sponsoring such a regional workshop, which could be hosted by FCh.

As innovative as this proposal might sound it also has precedent. In a letter of December 19, 1983, to Lic. Antonio Ortiz Mena (President of IDB), SOTEC explored the possibility of the IDB co-sponsoring a conference on Management of Technology Institutions in the Western Hemisphere. Attached to the letter was a preliminary list of 14 selected institutions that might be invited to participate in this event. No action was taken, partly because drafting a Plan of Operations in accordance with customary standards, including basic documents for the Workshop, proved beyond the means of the recently created SOTEC. In short, while the 1983 initiative was premature, a similar move would seem appropriate for the later 1980s.

An event of this sort would be a highly selective marketplace in which FCh could sell its unique management consulting skills to sister institutions and agencies. At the same time FCh could broaden its network of technology sources and reach out toward potential clients for its goods and services. A twinning association of FCh with a suitable sister institution in Chile or beyond might also result from such workshop contacts. After all FCh now has a comparative advantage in knowing how to plan, operate, and maintain the technology transfer processes. These skills tend to be underdeveloped in most of FCh's 300-odd sister institutions in Latin America and the Caribbean. In short, FCh could clone its ITT twinning with a fitting organization that has clearly defined its shortcomings in the management area and is already looking for remedial action. FCh staff members, who have direct hands-on experience in resolving management problems, could help their counterparts in the sister institution by bringing along copies of manuals, budgets, accounting systems, and maintenance programs used "at home," and assisting in their adaptation for use in the twin.

A carefully selected twin—with a managerial, political, legal and economic profile similar to FCh—could also become an effective penetration tool for selling FCh services to clients in the partner's home market.

With exception of the unique FCh-ITT relationship, I am not aware of any twinning among technology development and transfer institutions in Latin America. Yet most attributes of FCh indicate such twinning could overcome frequently encountered barriers between potential partners in Latin America and the Caribbean: minimal language and cultural differences; and proven experience with identifying projects, drawing up technical agreements, preparing contracts, and providing staff support in twin institutions. [4]

Additional potential windfalls from such a workshop might also include creation of a Latin American Society for Technology Development (LASTED).

It would be imprudent to venture a guess about how many of the above suggested actions would be implemented during the second decade of broadening FCh's outreach within Chile and beyond its borders. Yet I feel firmly that FCh is off the launching pad. Should it get successfully into orbit, the sky would be the limit to its potential.

By 1990 I would like to take another look at FCh's track record during its first 15 years of operations. In the meantime I will stay tuned for regular annual progress reports, which generous FCh management fortunately is making widely available.

NOTES

1.   In the United States the enterprise incubators are buildings in which a number of tenant-businesses share services. Aware of the high failure rate of enterprises in their early years, managers and owners of these incubators nurture their tenants with below-market rents, technical assistance, and business advice.

By 1985 there were over 120 industrial incubators in the United States, operating under varied sponsorships: (a) towns, aiming to create jobs; (b) universities and their professors, hoping to commercially exploit research and technology developed on campus; (c) private developers and venture capitalists wanting tenants for rehabilitated buildings in "old" commercial or industrial neighborhoods; and (d) private companies hoping to develop new business.

Several states have passed legislation to encourage creation of enterprise incubators. In 1984 Pennsylvania created a $17 million incubator loan program as part of a newly floated issue of state bonds.

Tenants in enterprise incubators are not immune to failure. Poor management, lack of capital, and flawed products kill potentially good projects. A study by the Small Business Administration (SBA) found that 70–80 percent of companies in U.S. incubators survive at least 3 years. This compares to only 30 percent of business created in the cold outside world.

2.   The following agencies accept DACON Form 1600:

         ADB   – Asian Development Bank
         ADFAED – Abu Dhabi Fund for Arab Economic Development
         BADEA   – Arab Fund for Economic Development in Africa
         IBRD   – World Bank
         IDB   – Inter-American Development Bank
         ILO   – International Labor Organization
         KFAED   – Kuwait Fund for Arab Economic Development
         UNDP   – United Nations Development Program

UNIDO   - United Nations Industrial Development
         Organization
WHO   - World Health Organization.

3.   The April 1981 report on the Role of the Inter-American
Development Bank in Latin America During the 1980s states: "The
Bank's participation in agricultural has suffered from chronic
shortage of investment projects. Traditional pre-investment
lending is not well-suited to agriculture because project identification and preparation is very complex, costly and time
consuming. As governments assign higher priority to agriculture,
attention to project identification and preparation becomes a
major concern" (p. 13).
    Likewise, Dr. William Demas, President of the Caribbean
Development Bank (CDB), in keynoting the Conference on "Opportunities for Small and Medium Sized Firms in Industry, Agriculture
and Trade" (held at the Organization of American States in
Washington, D.C, November 1, 1984), eloquently expressed the
problem in the following words: "Donor agencies find that
implementation of most of the large number of approved investment
and technical assistance projects are behind schedule. This is
because there is lack of absorptive capacity. The same difficul-
ty is encountered in formulation of new projects. And so, as
incredible as it might sound, the prime challenge is lack of well
prepared potential investment ventures, and not money. The
phenomenon is referred to as project drought, a problem encountered by many financial agencies."

4.   There is no need to reinvent the wheel for twinning. The
literature on the subject is growing by leaps and bounds. See:
Lauren Cooper, Twinning of Institutions: Its Use as a Technical
Assistance Delivery System. World Bank Technical Paper 23, 1984;
and Cities Explore Municipal Twinning, World Bank, The Urban
Edge, August 1985.

Epilogue

Chile is one of the more open economies in the world.
It would be pretty hard for me to find fault with the
way the economy is being managed.

Manager of a Chilean subsidiary
of a U.S. company [1]

In his eloquent foreword, Mr. Julio Luna described the
widely publicized controversies that surrounded the birth of the
Fundación Chile. ITT's Chilean subsidiary, the progenitor of
FCh, was just one of the many private enterprises nationalized
during the regime of President Allende. A short summary of what
happened to the Chilean economy in the last 14 years seems
therefore a good way to close this case study.

On September 11, 1973, when General Augusto Pinochet took
power in Chile, annual inflation was edging the 1000 percent
mark. Strikes were paralyzing transport and production, people
had to wait in long lines to buy such staples as bread and meat.

President Pinochet, following an economic program devised by
aides who studied at the University of Chicago (the "Chicago
school"), slashed tariffs, welcomed foreign investment, floated
domestic interest and foreign exchange rates, sold off hundreds
of state companies, and privatized the social security system.

The most far-reaching change has been privatization of the
more than 800 companies President Allende nationalized during the
1970-73 period. At the time of the 1973 coup, state-run
companies and public sector activities accounted for almost 40
percent of Chile's gross domestic product; by 1987 they accounted
for about 20 percent.

By mid-1987 the Chilean government had sold or was in the
process of selling all or parts of the state steel company, the

state airline, the national railroad, the state development bank, the government coal company, and had offered for sale to private investors 30 percent of ENDESA, the national electric company. Juan Antonio Guzman, the Education Minister, revealed plans to privatize the University of Chile's engineering school, as well as its hospital and television station.

Under the privatization scheme, the government sells—on the booming Chilean stock market—shares of state companies and institutes to local and foreign firms, to employees of nationalized companies, and to individual investors. More than 50,000 workers, or 1.3 percent of the entire labor force, have bought shares in privatized companies through this "grassroots capitalism" policy. Another 2.6 million employees belong to the privatized social security system.

CODELCO, the state copper mining company, is one of few enterprises the government is not willing to sell. Yet copper is far less important than it used to be. This is because the free-market policies have unleashed an agricultural export boom that has reduced copper's share of exports from 80 percent to about 40 percent. By 1988 Chilean farmers were exporting pears, apples, kiwi, grapes, and melons to the United States and Europe. Grape farmers have been so successful that U.S. growers have asked the Reagan administration to limit grape imports from Chile.

In 1987 Chile's foreign debt was $19 billion, the highest per-capita in Latin America, saddling the country with servicing payments of $2.2 billion, which wiped out the country's $1.1 billion trade surplus. Loans from multinational institutions had to cover most of the projected $1.1 billion current accounts deficit.

In 1985 Chile took the lead among Latin American nations in introducing so-called debt-equity swaps. Under this mechanism, both foreign and local companies buy debt of private and public sector firms from creditor banks at about a 30 percent discount. By mid-1987 Chile had converted $1.4 billion of debt into equity capital. General Pinochet aims to have the privatization and debt reduction policies firmly entrenched. If the Chilean people vote for return to democracy in the 1988 national plebiscite, a civilian government would presumably make only modest changes. This seems a pragmatic strategy, in that opposition leaders are nowadays willing to admit openly that "they would only partially roll back his [Pinochet's] policies if and when Chile returns to democracy."[1]

Shakespeare's Merchant of Venice counsels that "all that glitters is not gold." And so it is that the metamorphosis of Chile's economy from socialism to capitalism has not been painless. During the 1981-83 recession, one-third of the country's industrial companies had to be bailed out by their creditor banks. Many banks and financial institutions went bankrupt, and the central bank had to prop up dozens of others.

While the industrial and banking sectors seem to have recovered by 1988, millions of poor people are worse off. Unemployment in the shanty towns that ring Santiago has risen and consumption of basic foodstuffs has fallen. And there appear to be many violations of human rights.

Evidently the Chilean coins—and all the rest of the coins of the world—have at least two sides, both of which might not be equally pretty. The beauty will obviously depend on the eye of the beholder.

NOTE

1.   Quoted from Tyler Bridges, "Pinochet's Reforms Boost Chilean Economy," Washington Post, September 8, 1987, H6.

Appendixes

1976:
1. Feasibility study for increasing nutritional value of bread and noodles.
2. Formulation of a cracker, high in calories and protein, for use in the National School Lunch Program.
3. Improvement program for cereals, legumes and oil bearing crops with high oil and protein content.

1977:
4. Organization of first international seminar aimed at disseminating market and economic information, related to attractive investment opportunities in Chile.
5. Study of lupine utilization in human nutrition.
6. Purchase of 5000 m^2 in the Recabarren-Manquehue Park for construction of FCh headquarters with 1600 m^2 of a pilot processing plant, 700 m^2 of laboratory space, and 2000 m^2 of offices.
7. Establishment of a Technical Assistance Center (CAT) for the canning industry.

1978:
8. Construction of FCh headquarters initiated.
9. Introduction of systems for handling and storage of fresh fruits and vegetables.
10. Planning of rural telephone networks.
11. Adaptation of technology for more effective catch and processing of marine resources.

12. Establishment of a mariculture center in Coquimbo for production of algae under conditions of controlled environment.
13. Making laboratory services available for third parties.

1979:
14. Initiation of quality control program via a seminar for staff of the canning industry.
15. Program for improved post-harvest handling of fresh fruits and vegetables under joint auspices of producers and marketers.
16. Second international seminar for promotion of investment opportunities in Chile.
17. Inauguration of FCh headquarters.
18. Vocational training for dairy farmers in Region 10, jointly organized with the National Television of Chile (TNC) and National Training Institute (INACAP).
19. Initiation of technical assistance to growers of asparagus, aimed at promoting off-season exports to markets in the Northern Hemisphere.
20. Consultancy under CAT auspices, to agroindustries in energy utilization and food dehydration technology.

1980:
21. National technical assistance program for improvement of small cheese factories.
22. Initiation of salmon breeding work at Curaco de Velez, the first of four locations of the Salmones Antártica subsidiary enterprise.
23. Initiation of oyster cultivation in Coquimbo.
24. FCh library converted into a technical information center.
25. FCh's food laboratory sets up organoleptic and sensory panels.
26. First applications of microprocessors to establishment of: (a) private telex system for the El Teniente mine of the Chilean Copper Corporation (CODELCO); and (b) control of refrigeration system for the Concón Refinery of Chilean Petroleum Corporation (ENAP).
27. Development of new dairy products (Swiss-style yogurt, dairy desserts) and adaptation of ultra high yemperature (UHT) processing of milk, aimed at extending shelf life.
28. Installation of panels on the roof of the FCh headquarters to test industrial application of solar energy.

1981:
29. Video-tape on FCh field work prepared for purposes of disseminating understanding of concepts related to transfer of technology.
30. Precooking and freezing technology introduced for processing of shellfish for the local Chilean markets.

31. Commercial operations initiated for processing of fruit pulp for use in yogurts, to substitute for previously imported products.

32. Technical assistance to asparagus growers aimed at increasing planted acreage from 200 to 800 ha, and pilot export shipments to markets in the Northern Hemisphere.

33. Importations of salmon eggs from the United States to gain experience with breeding this valuable species in southern Chile.

34. New equipment in the pilot processing plant to facilitate formulation of new foods for use by local industries.

35. Two rural telephony projects initiated in Region 10, introducing to Chile subscriber carrier technology with buried cables.

36. Microprocessor unit completed first installations in mining and energy enterprises.

37. Intensive public relations and promotion campaigns aimed at increasing demand for FCh services.

38. The Agricultural Service (SAG) authorized FCh to serve as official laboratory to measure pesticide residues in fruits and vegetables, using gas chromotography, and issue corresponding export certificates.

1983:

39. Developed food products based on Pisco Capel.

40. Appearance of first Technical Publication, issued by FCh and written by Chilean as well as foreign experts.

41. First dietary product released: rice flower for consumers unable to eat wheat flower.

42. Inauguration of FCh Mariculture Center at Tongoy for production of Pacific oyster (Crassostrea gigas).

43. Fruit quality control program certified 7.4 million boxes, up from 4.5 million cases in 1981/82.

44. First forestry project established in Puerto Carmen, at the southern-most edge of the Great Chiloé island.

45. Completed studies of technical and economic feasibility of the jojoba bean bush, and initiated production of 15,000 plants from certified seed.

46. Initiated formulation of exotic cheese such as camenbert, brie, and emmenthaler.

47. Issued second revised edition of FCh promotion booklet, and a videocassette for all FCh services.

48. Seminar on production and marketing of asparagus.

49. Land acquired at Chiloé for establishment of a new salmon hatching station (Astilleros), using eggs imported from the United States as well as eggs from first returns of salmon released at Curaco de Vélez.

1983:

50. Fresh fish introduced to supermarkets under FCh brand name "Don Pez" (Mr. Fish) to demonstrate feasibility of producing and mass marketing quality sea food.

51. Development of cheeses adapted to characteristics of local dairy industry and capable to substitute for importations.

52. Study of red meat marketing leads to creation of PROCARNE, a FCh subsidiary enterprise specializing in production of vacuum packed boxed beef.

53. Publication of Agro-Economic Information (IAE) bulletin initiated in Region 4. IAE aims to guide farmers in their decision making based on data on crop profitability, demand, and price outlook.

54. Initiation of pilot project focused on establishing feasibility of using wood-intensive housing in Chile.

55. A 3-year marketing consultancy contract signed with the University of Santa Maria aimed at establishing a student recruitment program among high school graduates.

56. Four technical publications issued: growth regulators for apples and pears, plum production in California, production of table grapes in the United States, and alternative technologies for processing rice.

57. CAPRILAC pasteurized goat cheese processing plant constructed in Ovalle (Region 4) to help improve income and quality of life among low-income goat herders of the region.

58. FCh quality program certifies 10.5 million boxes of fresh export fruit, or 25 percent of the total of 42.0 million boxes exported from Chile.

59. First salmon fingerlings liberated in the Pacific Ocean near Curaco de Vélez.

60. Pilot food processing plant formulates several types of camenbert and submits them for evaluation by FCh tasting panels.

61. Method for smoking of salmon developed, and resulting product evaluated by FCh tasting panel.

62. Salmones Antártica created as FCh subsidiary, consolidating all activities related to salmon breeding.

1984:

63. Several technical and commercial missions visit Chile to explore potential investment in forestry. Factory established for production of prefabricated wood-intensive housing.

64. Two Region 10 Agro-Economic Information bulletins published.

65. PROCARNE vacuum packed boxed-beef plant goes on stream.

66. Program for technology transfer for wine and alcoholic beverages initiated, focusing primarily on improvement of quality.

67. Growth of salmon production at Salmones Antártica makes it necessary to develop suitable packages and internal distribution channels.

68. Open sea salmon ranching (a mar abierto) established at Astilleros (Region 10) and Rio Prat (Region 12), increasing production capacity of Salmones Antártica four-fold.
69. Pilot project initiated for use of computers in management of dairy herds belonging to members of the Agricultural Cooperation in Frutillar (CAFRA).
70. Pilot plant experiments with processing of waste from fruit and wood.
71. Major marketing promotion campaign initiated for the Tongoy brand of oysters produced by Cultivos Marinos (CULTIMAR), an FCh subsidiary enterprise.
72. Adaptation of production, handling and processing of kiwi fruit introduced to Chile from New Zealand.
73. Organization of 4 seminars aimed at diffusion of location-specific forestry technologies.
74. Mass media oriented public relations campaign initiated aimed at creating an image for FCh as center for technology transfer.
75. Three types of Chevrita branded goat cheeses released for sale: molded, camenbert, and creamy.

1985:
76. Tongoy oyster breeding station produces 3 million eggs, or about 25 percent of total natural oyster production in Chile. Some 2 million eggs delivered to 22 contract oyster family producers in Region 10 to grow out the Pacific oyster for distribution by FCh.
77. Program for Transfer of Technology in Wine and Alcohol signed quality control contracts with major Chilean wineries and with a large producer of pisco in Region 4.
78. Quality control and certification program expanded from fresh asparagus, apples, and table grapes to include nectarines, plums, melons, and onions.
79. Quality control program extended to canned fruit.
80. SIC-15 multi-purpose computer system established for use in industrial automation and control processes.
81. FCh issues a manual for establishment and management of radiata pine nurseries.
82. Two FCh technical reports issued: Technology for Production of Chilean Wines, and European Market Potential for Fresh Vegetables.
83. PROCARNE plant officially initiated with daily capacity for processing 140 head of cattle. FCh signed a contract to provide technical and marketing advice to the new subsidiary.
84. PROCARNE vacuum packed boxed-meat accepted by retailers, supermarkets, and butchers; traditional systems of marketing red meat modified accordingly.
85. FCh's Program of quality control certification and for fresh fruit covers one-fourth of Chilean exports.

86. Some 17,000 copies of the semi-annual Agro-Economic Information Bulletin are distributed, indicating its acceptance by farmers.

NOTE

1. Based on Annex to the <u>Fundación Chile 1985 Annual Report</u>.

APPENDIX 2
LOCATION OF SALMON FARMS AND RANCHES IN CHILE, 1983

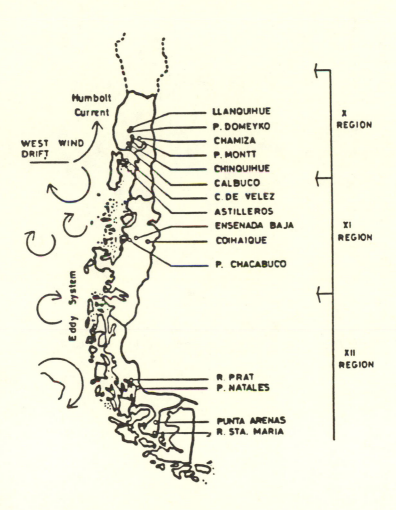

Source: Carlos Wurmann Gotfrit, "Chilean Salmon: From Dreams to Reality," Infofish Marketing Digest, April 1985.

REGIONAL DISTRIBUTION OF PROJECTS, FUNDACION CHILE, 1983

Location b/	Type of Project a/										
	1	2	3	4	5	6	7	8	9	10	11
1. Arica		X				X					
2. Iquique											
3. Chuquicamata									X		
4. El Salvador									X		
5. Copiapó	X	X				X					X
6. Vallenar	X	X									
7. La Serena	X	X	X								
8. Tongoy						X	X				
9. Ovalle	X		X	X							
10. San Felipe	X	X	X								X
11. Con Cón											
12. Los Andes	X	X									
13. Valparaíso	X			X							X
14. Llay-Llay	X			X							
15. San Antonio	X										
16. Santiago	X		X	X					X	X	X
17. Buin	X	X									
18. Rancagua	X	X							X		
19. San Fernando	X			X							
20. Curicó	X	X	X								
21. San Javier		X									
22. Constitución											
23. Linares	X	X		X							
24. Parral		X	X								
25. Chillán	X			X							
26. Concepción				X							X
27. Laja											
28. Los Angeles				X							
29. Temuco											
30. Valdivia											
31. Osorno					X					X	
32. Puerto Varas				X						X	
33. Castro						X		X			
34. Quellón											X
35. Coihaique		X								X	
36. Puerto Chacabuco						X		X			
37. Puerto Natales						X		X			
38. Punta Arenas						X		X	X	X	

Source: Fundación Chile

a/ Code: 1/ Quality Control for Export Fruit; 2/ Asparagus, berries and other crops; 3/ Agroindustries; 4/ Dairy and Caprilac; 5/ Procarne; 6/ Sweet Water Fish Farming; 7. Marine Fish Farming; 8/ Salmon Production; 9/ Microprocessor; 10/ Rural Telephone; 11/ Jojoba and Forestry.

b/ Listed from North to South

Bibliography

CHILE

Public Sector

The most comprehensive public statistical sources are: Instituto Nacional de Estadística (INE), the Central Bank, and the Planning Office (Odeplan).

Other government agencies and companies publish specialized data: Bank Superintendency on interest rates, return on financial instruments, and legal reserve requirements; Chilean Copper Commission releases copper prices and production figures; Foreign Investment Committee supplies limited information on authorized and actual foreign investments in Chile.

The public sector data are generally reliable. However, information is usually not dated, and agencies tend to be secretive and sometimes refuse to furnish data even if they are available.

Many state agencies release their data irregularly or alter methods in which data are prepared; this makes comparisons over time difficult for banking, export, foreign debt, and foreign investment data. Local observers advise caution with INE statistics on unemployment and consumer and wholesale price indexes.

Private Sector

Dailies and weeklies are a good way to track economic and political events. To get a balanced picture, a variety of newspapers and magazines must be read, because each publication emphasizes different aspects of the same story. It is useful to read the private dailies El Mercurio, La Tercera, La Segunda, and the public sector paper La Nación, plus the weekly business magazine Estrategia, all politically conservative.

Due to censorship by the Dirección Nacional de Comunicación Social (Dinacos), political events and social climate are hard to monitor. Reporters stick to official versions of events.

Opposition papers are usually weeklies: Hoy, associated with the Christian Democrats, is informative, respected, and widely read. Análisis is an opposition weekly, connected with the Socialist Party's MDP faction; it has been shut down, its director arrested and released.

The Catholic Church also publishes news and commentary on economic and political events in Chile. The monthly Mensaje reports on Christian thought about a wide spectrum of topics. The Vicaria de la Solidaridad publishes a weekly, which documents human rights violations and ways economic events affect the working class in various industries.

Among the most useful sources for business and economic information are bulletins published by chambers of commerce and industrial and agricultural associations. The Chilean and Santiago Chambers of Commerce provide daily coverage on taxes, customs duties, maritime transport, exports and imports, plus a variety of data reprinted from official sources.

The Sociedad de Fomento Fabril (SOFOFA) periodically publishes an industrial production index, whose figures often differ from those reported by the INE. Mercado y Opinión Pública supplies retail sales indexes.

Agricultural data can be obtained from the Sociedad Nacional de Agricultura, and the Fundación Chile, which covers only administrative regions 6, 7, and 10. The Universidad de Chile publishes unemployment data that are generally higher than official estimates. The Universidad Católica de Chile's Business and Economics Department puts out an economic bulletin, which tends to advocates doctrines of the Chicago School of Economics.

The New York branch office of the Production Development Corporation (CORFO) makes available, free of charge, monthly Chile Economic Report (CORFO, Suite 5151, One World Trade Center, New York, NY 10048). See also Chile country reports by the IDB, the IMF, and the World Bank.

The Fund of Multinational Management Education (FMME), Lagniappe Letter and North American-Chilean Chamber of Commerce co-sponsor annual Analysis and Outlook sessions aimed at assessing economic and political issues and their implications for international business executives interested in Chile.

CHILE FOUNDATION

Articles about FCh Published Abroad

1. Robert H. Cotton and Steward S. Flaschen. "Foundation Chile—A New Strategy for Industrial Development," Cereal Foods World Vol. 23 No. 7 (1978): 368-370.
2. Frank C.H. Wylie. "The Chile Foundation," The Review of the River Plate (Buenos Aires), No. 4119 (September 19, 1980): 442.
3. Theodore Kendall. "Chile's Food Industry Responds to Technological Development," Food Development 15 No. 6 1981): 24-27.
4. Anon. "The Growing Importance of Chilean Fruits," International Fruit World 2 (1982): 2983-2992.
5. Anon. "Quality Control for Chilean Fruit," International Fruit World 3 (1982).
6. Anon. "Chilean (Rice Extension) Plant Commences Production," LEG Newsletter (Colorado State University) 1 No. 1 (1983): 3.
7. Peter Brown. "Chile Joins Salmon Farmers," Fish Farming International 10 No. 8 (1983): 8-11.
8. George Giddings. "South American Hakes: Resource and its Utilization," Marine Fisheries Review 42 No. 1 (1980): 8-11.
9. George Giddings and Matoira M. Chanley. "Developments in Mass Culture of Brine Shrimp." In: Advances in Food Producing Systems for Arid and Semiarid Lands, J.T. Manassah and E.J. Briskey (eds.). New York: Academic Press, 1981; pp. 1021-1051.
10. Pablo Herrera. "La Experiencia de Pulpa de Pescado en Chile" (Experience with Fish Pulp in Chile), Non-Traditional Fish Products for Massive Human Consumption, Final Report on the Round Table held in Washington, D.C.: Inter-American Development Bank, 1981.
11. J.J. Romero. "Production and Use of Lupines in Chile," Agricultural and Nutritional Aspects of Lupines, proceedings of First International Lupine Workshop, held on April 12-21, 1980 in Lima-Cuzco (Peru). Eschborn: German Agency for Technical Cooperation (GTZ), 1981, pp. 806-817.
12. Fernando Sánchez. "Producción e Industrialización de Granos en América Latina" (Grain Production and Industrialization in Latin America), Ciencia Interamericana 20 Nos. 3-4 (1980).
13. S.S. Flaschen and R.H. Cotton. "ITT Transfers Food Technology to Chile," Milling and Baking (June 24, 1980); 39-40.

Articles about FCh published in Chile

1. Universidad del Norte-Centro de Investigaciones Submarinas (UN-CIS), Proceedings of International Symposium on Coastal Upwelling. Santiago de Chile: FCh, 1976.
2. Sanidad e Higiene en la Industria Procesadora de Alimentos (Sanitation and Hygiene in the Food Processing Industry). Santiago de Chile; FCh, 1976.
3. La Calidad como Factor de Desarrollo de la Industria Conservera—I (Quality as a Development Factor in the Canning Industry). Santiago de Chile: FCh, 1977.
4. La Calidad como Factor de Desarrollo de la Industria Conservera—II (Quality as a Development Factor in the Canning Industry). Santiago de Chile: FCh, 1978.
5. Calidad en la Industria Conservera—III (Quality in the Canning Industry). Santiago de Chile: FCh, 1979.
6. Situación, Análisis y Perspectivas del Lupino en Chile (Status, Analysis and Outlook for Lupin in Chile). Santiago de Chile: FCh, 1978.
7. Chile y su Industria Pesquera (Chilean Fishing Industry), Santiago de Chile: FCh and PROCHILE, 1980.
8. Hermann Schmidt Hebbel. Aditivos y Contaminantes de Alimentos—Reglamentación de Alimentos (Additives and Contaminants—Regulation of Foods). Santiago de Chile: FCh, 1979.
9. Hermann Schmidt Hebbel. Las Especias-Condimentos Vegetales (Spices and Plant Condiments). Santiago de Chile: FCh, 1980.
10. Hermann Schmidt Hebbel and I. Pennacchiotti. Las Enzimas en los Alimentos (Enzymes in Foods). Santiago de Chile: FCh, 1982.
11. El Desarrollo de la Industria Pesquera: Análisis de Decisiones de Inversión y Tecnología en el Sector (Development of the Fishery Industry: Analysis of Technology and Investment Decisions in the Sector)" Santiago de Chile: FCh and Subsecretaría de Pesca, 1981.
12. Boletines del Centro de Asistencia Tecnológica (CAT) de FCh, 1978-1979.
    a. Economía de Energía en la Industria Conservera (Energy Economics in the Canning Industry);
    b. El ABC de las Buenas Prácticas de Elaboración (The ABCs of Good Manufacturing Practice);
    c. Guías para Instrucción de Operadores de Autoclaves (Guides for Training Autoclave Operators);
    d. Pelado Químico de Papayas (Chemical Peeling of Papayas);
    e. Principios de Estirilización Comercial en Conservas (Principles of Commercial Sterilization in Canned Foods);
    f. Autoclaves para Esterilización de Conservas (Autoclaves for Sterilization of Canned Foods).

13. El Papel de la Tecnología en una Estrategia de Desarrollo (The Role of Technology within Development Strategy). Seminario Internacional organizado por la Sociedad Chilena de Tecnología para el Desarrollo, Abril 1982, pp. 16.

14. Manuel Elgueta G. and Eduardo Venezian L. (eds.). Economía y Organización de la Investigación Agropecuaria (Economics and Organization of Agricultural Research). Proceedings of a Seminar held in Santiago de Chile, May 1979, pp. 297.

15. La Industria de Cereales y sus Alcances Technológicos (The Cereals Industry and its Technological Achievements). Santiago de Chile: FCh and Universidad Técnica del Estado, 1979.

16. División Frutas y Hortalizas de FCh, Publicaciones Técnicas (traducciones), 1982:
    a. Forshey, C.G., and D.C. Elfving. Poda de Formación del Manzano (Pruning of Apple Trees). Ithaca, NY: Cornell University, Division of Agricultural Science;
    b. Aplicación de Nutrientes y Plaguicidas por Sistema de Reigo por Goteo (Applying Nutrients and Herbicides by Drip Irrigated). University of California: Division of Agricultural Sciences;
    c. La Rue, J.H., and M.H. Gerdts. Cultivo de Ciruelo en California (Commercial Plum Growing in California). University of California: Division of Agricultural Sciences.

17. Alfred Vial C. "Rentabilidad de la Producción e Industrialización de la Leche de Cabra" (Profitability and Production of Goat Milk), Tecnología y Agricultura (Jan-Feb. 1982).

18. Pablo Espinosa and Alfredo Vial C. "Curso de Comercialización y Procesamiento de Productos Caprinos" (Course in Marketing and Processing of Goat Products), Universidad de Chile Ovalle (Sept. 1983).

19. Informativo Agro-Económico - VI Región (Agro-Economic Information—Region 4), May 1983.

20. Plan Operacional (Operations Plan), 1980, 1981, 1982, 1983, 1984, 1985.

21. John Drew. "Aprovechamiento de Subproductos de Celulosa" (Utilization of Pulp By-Products), Chile Forestal (Nov. 1982).

22. "Impactante Crecimiento del Cultivo del Espárrago" (Substantial Increase in Asparagus Growing), Tecnología y Agricultura (June/July 1984).

23. M. Wayne Sandvig. Report to Board of Directors of Fundación Chile, January 18, 1984, pp. 4.

24. Gustavo Chian. "Queremos Ayudar y no Competir con la Empresa Privada" (We Want to Help, Not Compete with Private Enterprise), Chile Forestal (Dec. 1983): 6-7.

25. Andres Saiz. "Investigación y Tecnología para Mundo Mejor" (Research and Technology for a Better World), Revista Munda (enero 1984): 56-59.

26. Juan Pablo Torrealba. "El Potencial de la Jojoba en Chile" (Jojoba Potential in Chile), Próxima Década (March 1984): 16-20.
27. María Isabel Diez. "Wayne Sandvig—Un Perfil," La Segunda (April 6): 1984.
28. "Países Latinoamericanos Enfrentan en Conjunto Problemas de Calidad en Productos del Mar" (Latin American Countries Jointly Face Quality Challenges of Marine Products), Chile Pesquero (March 1984): 42-43.
29. Marvin H. Gerdts and James H. La Rue. "Todo lo que Usted Desea Saber sobre el Ciruelo" (All You Want to Know about Plums), Chile Agrícola (June 1984): 151-154.
30. Fernando Sánchez and James A. Duke. "La Papa de Nadi" (Potatoes from Nadi), El Campesino (April 1984): 15-17.
31. "La Heróica Lucha del Vino" (Heroic Struggle of Wine), El Mercurio—Revista del Campo (June 18, 1984): 11-13.
32. "Rentabilidad de Cultivos en el Riego de la VII Región Temporada 1984-1985" (Profitability of Crops under Irrigation in Region 7 during the 1984-1985 Season), El Mercurio—Revista del Campo (May 21, 1984).
33. "Irradiar la Fruta Fresca de Exportación?" (Irradiation of Fresh Export Fruit?), Tecnología y Agricultura (June/July 1984): 38.
34. Pablo Herrera L. "Irradiación de Productos del Mar" (Irradiation of Sea Foods), Revista Ingenieros (June 1984): 19-21.
35. Wayne M. Sandvig. "La Tecnología es la Base para el Desarrollo del País" (Technology as Basis for Country Development), Gestión (Nov. 1984).
36. Gonzalo Melo Gecele. "Sistemas Microcomputacionales de Control Desarrollados para la Industria Nacional," (Microcomputation Systems Controls Applied to national Industries), Revista Ingenieros (June 1984).
37. Gonzalo Melo Gecele. "Sistemas de Control Aplicados a Procesos Complejos de Cobre y su Evolución," Minería Chilena (Aug. 1984).
38. Herman Eyzaguirre. "Hacia una Nueva Cocina a la Chilena" (Toward a New Cooking a la Chilena), Mundo (Sept. 22, 1984).
39. Rosario Valdes Chadwick. "Fundación Chile: Un Reto al Paladar Chileno" (A Guide to Chilean Taste), Paula (Nov. 1984).
40. "Ulex Europea—al Acecho de Nuestros Bosques" (European Ulex-Menace to Our Forests), Tecnología y Agricultura 4 (1984): 30-31.
41. Barbara Delano. "La Ostra Japonesa en Chile" (The Japanese Oyster in Chile) Informativo de la Cámara de Comercio de Santiago (July 2, 1985).
42. Peter Brown and Ricardo Rodrigues. "Ranching of Pacific Salmon in Southern Chile: A Progress Report on Salmones Antártica." Paper presented at International Conference on Biology of Pacific Salmon, Victoria (B.C.) Canada, September 1984.

43. Peter Brown. "Perspectivas de la Introducción de Salmones en la Zona Austral de Chile" (Outlook for Salmon Introduction to Southern Chile), Informativo Aysen (October 1984).

44. Patricia Salinas. "Aumento de Producción de Ostras con Cultivo Artesanal" (Increase in Artisanal Production of Oysters), La Tercera Regional (Nov. 23, 1984).

45. Raúl Contuarías. "La Voraz Polydora" (The Voracious Polydora), Hoy (Dec. 2, 1985).

46. "Aprovechando Residuos de Mataderos y Nueva Dietas para Animales a partir de Residuos" (Utilizing Slaughterhouse By-products and Livestock Feeds Based on Crop Waste), El Tartersal (Dec. 19840.

47. "Buenas Expectivas para Frutales Menores" (Good Outlook for Minor Fruits), El Campesino (Aug. 1984).

48. F. Sánchez A. "Agroindustria Hortofruticola Nacional—Potencial y Perspectivas de Desarrollo (Produce Based Agro-industries—Potential and Outlook for Development), El Campesino (Oct. 1984).

49. Frederick Jensen. "El Desgrane en la Uva de Mesa: Causas y Prevención" (Shatter Problem of Thompson Seedless Table Grapes—Causes and Prevention), Tecnología y Agricultura (Aug./Sep. 1984).

50. Mauricio Meyer. "Capacidad Instalada de Frigoríficos par Fruta a Nivel Nacional" (National Refrigeration Capacity for Fruit), FCh (1984).

51. David de Curto. "Vanguardia de Pioneros en Exportaciones de Frutas Frescas" (Leader Among Pioneer Exporters of Fresh Fruit), FCh (Dec. 11, 1984).

52. "Exportación de Hortalizas Frescas con potential de Mercado en Europa" (Export Potential for Fresh Vegetables in European Markets), FCh (Nov. 1984).

53. "Hortalizas Con Futuro en Europa" (Vegetables "with future" in Europe), El Mercurio—Revista del Campo (Nov. 19, 1984).

54. "Desarrollo de Instrumentos para la Transferencia Tecnológica en Producción de Leche" (Development of Tools for Technology Transfer in Milk Production), FCh (1984).

55. "Frutillar—Computación Llega a Establos Chilenos del Sur" (Frutillar - Computation Arrives to Stables in Southern Chile), La Tercera (Aug. 23, 1984).

56. CAFRA. "50 Años al Servicio de la Agricultura" (50 Years in Service to Agriculture), Agro-Análisis (Sept. 1984).

57. "Queso de Cabra—del Comercio Clandestino a un Producto Ejemplar" (Goat Cheese—from Clandestine Commerce to a Model Product), Tecnología y Agricultura (Aug./Sept. 1984).

58. "Atmosfera Modificada—Supra Percibilidad de Productos Hortofrutículas" (Modified Atmosphere Overcomes Perishability of Fresh Produce), Estrategia (Nov. 5, 1984).

59. "Desarrollo de Ventajas Comparativas de la Madera en la Construcción de Vivienda" (Development of Comparative Advantages of Wood in Housing Construction), FCh—University of Chile (Faculty of Architecture and Urban Development,

Schools of Agricultural, Veterinary and Forestry, Faculty of Physical and Mathematical Sciences), 1983.
60. "Dátiles—La Novedad de un Fruto Milenario" (Dates—The New Thousand Year Old Fruit), Estrategia (Oct. 1-7, 1984).
61. Cornelius Foote, Jr. "No Sour Grapes in Chile," Miami Herald (April 12, 1985).
62. "Elevar La Calidad de Nuestros Vinos" (Improve Quality of Our Wines), Tecnología y Agricultura (Aug./Sept. 1984).
63. Sociedad Chilena de Tecnología para el Desarrollo (SOTEC):
    1. Premio Anual de la Innovación, Oct. 1983;
    2. Reunión Anual de SOTEC, Nov. 1983;
    3. Cuenta del Directorio, Nov. 1984;
    4. Socios de SOTE al 22 de marzo del 1985.
64. "Sistema de Carne en Caja" (Boxed Beef Systems), Tecnología y Agricultura 4 (1984).
65. Emilio Castro C. "El Uso de Cultivos Starters en la Industria Cecinera" (Use of Culture Starters in Sausage Making Industries), Revista Ingenieros (June 1984).
66. Pablo Herrera L. "Irradiación de Productos del Mar," (Irradiation of Seafoods), Revista Ingenieros (June 1984).
67. "Proceedings of International Seminar on Salmon Farming Perspectives in Chile." Fundación Chile (March 17-19, 1987).

Promotional Folders

1. Memoria (Annual Reports), 1976-84.
2. Fuente de Avanzada Tecnología entre Chile y el Mundo (Source of Advancing Technology between Chile and the World), 2 folders.
3. Laboratorio (Laboratory).
4. Todo un Universo Inteligente (The Microprocessors Universe).
5. Programa Control de Calidad—Frutas y Hortalizas (Quality Control Program—Fruits and Vegetables).

SELECTED WRITINGS ON TECHNOLOGY TRANSFER

Articles and Periodicals

1. Robert H. Cotton and Steward S. Flaschen. "Foundation Chile—A New Strategy for Industrial Development," Cereal Foods World (July 1978).
2. Robert H. Cotton, and Steward S. Flaschen. Technology that Works for Developing Nations, World Futures Studies Conference, West Berlin, May 1979.
3. Sol Divita (ed.). Marketing New Technologies. Papers presented to meeting of Metropolitan Washington Chapter of the American Marketing Association, held at Marriott Hotel, Crystal City, Arlington, Virginia, November 8, 1984.

4. Alice B. Lentz (ed.). U.S. Cooperation in Science and Technology for Development in the ASAEN Nations. New York: Fund for Multinational Management Education (FMME), October 9, 1980.
5. Edwar Pilgrim and Frank Meissner. "Transfer of Intermediate and Light- Capital Food Technologies in Latin America," Journal of the International Appropriate Technology Association (March-June 1984): 16-31.
6. Harvey W. Wallender. "Developing Countries Orientations Toward Foreign Technology in the Eighties: Implications for New Negotiation Approaches," Columbia Journal of World Business (Summer 1980).

BOOKS

1. Anon. Technology, Innovation and Regional Economic Development, Washington, D.C.: Office of Technology Assessment (OTA), U.S. Congress, 1984, 167p.
2. Anon. Science and Technology Indicators: Resources Devoted to R&D. Paris: Organization for Economic Cooperation and Development, 1984.
3. Jack N. Behrman and William A. Fischer. Science and Technology for Development. Cambridge, MA: Oelgeschlager, Gunn & Hain, 1976.
4. Jack N. Behrman and Harvey W. Wallender. Transfers of Manufacturing Technology within Multinational Enterprises. Cambridge, MA: Ballinger, 1976.
5. William Beranek, Jr. and Gustav Ranis. Science, Technology and Economic Development. New York: Praeger, 1978.
6. Domien H. Bruinsma, Wouter W. Witsenburg, and Willem Wurdemann. Selection of Technology for Food Processing in Developing Countries. Waeningen: Netherlands Agricultural University, 1983, p. 212.
7. Daniel and Masafami Nagau. Capital Goods Production in the Third World: an Economic Study of Technology Acquisition. New York: St. Martin's Press, 1983.
8. Wonter Van Ginneken and Christopher Baron (ed.). Appropriate Products, Employment and Technology: Case-studies on Consumer Choice and Basic Needs in Developing Countries. New York: St. Martin's Press, 1984, 260 p.
9. Stanislaw Gomulka and Alec Nove. East-West Technology Transfer. Paris: Organisation for Economic Cooperation and Development (OECD), 1984, 94 p.
10. Denis Goulet. The Uncertain Promise: Value Conflicts in Technology Transfer. New York: IDOC/North America, 1977.
11. Carl J. Dahlman, Bruce Ross-Larson, and Larry E. Westphal. Managing Technological Development, Washington, D.C.: World Bank Staff Working Paper, 1984, 54 p.

12. Joseph J. McInar and Howard A. Clonts (eds.). Transfering Food Production Technology to Developing Nations: Economic and Social Dimensions. Boulder, CO.: Westview, 1983.

13. National Research Council. Technology, Trade and the U.S. Economy. Washington, D.C.: National Academy of Sciences, 1976.

14. Silvere Seurat. Technology Transfer: a Realistic Approach. Houston, TX: Gulk, 1979.

15. Frances Stewart. International Transfer of Technology. Washington, D.C.: World Bank Staff Working Paper 344, 1979 (second printing, 1985), 116 p.

16. Frances Stewart and Jeffrey James (eds.). The Economics of New Technology in Developing Countries. London: Frances Pinter, 1982.

17. Robert Stobaugh and Louis T. Wells, Jr. (eds.). Technology Crossing Borders: Choice, Transfer and Management of International Technology Flows. Cambridge, MA: Harvard Business School Research Colloquium Series, 1984, 329 p.

18. Moshe Syrquin and Simon Teitel (eds.). Trade Stability, Technology, and Equity in Latin America. New York: Academic Press, 1982.

19. Nancy S. Truitt and David H. Blake. Opinion Leaders and Private Investment—An Attitude Survey in Chile and Venezuela. New York: Fund for Multinational Management Education, 1976, p. 56.

20. Harvey W. Wallender. Technology Transfer and Management in Developing Countries. Cambridge, MA: Ballinger, 1979.

21. Charles Weiss and Nicholas Jaquier (eds.). Technology Finance and Development. Lexington, MA: Heath, 1984, 343p.

22. Martin Fransman and Kenneth King (eds.). Technological Capability in the Third World. London: MacMillan, 1984, 404p.

23. UNIPUB (205 East 42 St, New York, NY 10017):
    1. Anon. Science and Technology Education and National Development. 1984, 197 p.
    2. A. Bhalia, D. James, and U. Stevens. Blending New and Traditional Technologies, 1984, 285 pp.
    3. Hyung Sup Choi. Bases for Science and Technology Promotion in Developing Countries, 1984, 295 p.
    4. Maurice Goldsmith, Alexander King, and Pierre Laconte (eds.). Science and Technology for Development—The Non-Governmental Approach, 1984, 194p.
    5. Alan Hancock (ed.). Technology Transfer and Communication, UNESCO, 1985, 385p.
    6. Barbara G. Lucas and Stephan Freedman (eds.). Technology Choice and Change in Developing Countries—Internal and External Constraints. 1983, 155p.
    7. Hans Singer. Technologies for Basic Needs. Geneva: ILO, 1982, 161p.

8. Earnst U. Von Weizacker, M.S. Swaminathan, and Aklilu
    Lemma (eds.). New Frontiers in Technology Application:
    Integration of Emerging and Traditional Technologies.
    1983, 272p.

24. United Nations Industrial Development Organization (UNIDO).
     Development and Transfer of Technologies Series:
     1. National Approaches to the Acquisition of Technology;
     2. UNIDO Abstracts on Technology Transfer;
     3. The Manufacture of Low-cost Vehicles in Developing
         Countries;
     4. Manual on Instrumentation and Quality Control in the
         Textile Industry;
     5. Technology for Solar Energy Utilization;
     6. Audio-visual Techniques for Industry;
     7. Technologies from Developing Countries;
     8. Process Technologies for Phosphate Fertilizers;
     9. Process Technologies for Nitrogen Fertilizers;
     10. Brickmaking Plant: Industry Profile;
     11. Technological Profiles on the Iron and Steel Industry;
     12. Guidelines for Evaluation of Transfer of Technology
          Agreements;
     13. Fertilizer Manual;
     14. Case-Studies in the Acquisition of Technology;
     15. Technological Self-Reliance of the Developing
          Countries: Towards Operational Strategies.

Index

About the Author

Born in Czechoslovakia in 1923, Dr. Meissner served in the Royal Air Force (RAF) in England during World War II; in 1949 he obtained a B.Sc. in Agronomy from the Royal Agricultural University in Copenhagen; in 1950 a M.S. in Agricultural Economics from Iowa State University in Ames; in 1957 a Ph.D. in Marketing from Cornell University at Ithaca (N.Y.); and in 1965 attended the International Marketing Institute (IMI), held at the Harvard University Graduate School of Business.

During 1969-1988, Dr. Meissner was a marketing economist in the Agriculture and Forestry Development Division of the Inter-American Development Bank (IDB) in Washington, D.C., engaged in identification, selection, preparation, analysis, implementation and monitoring of some 75 agricultural and agroindustrial projects in over 20 countries of Latin America and the Caribbean.

Dr. Meissner has published over 100 papers and articles in professional and trade journals; co-edited three books (International Business Management, in 1964; Marketing Systems in Developing Countries, in 1976; and Agricultural Marketing in Developing Countries, in 1976). Since 1987 Dr. Meissner has been Adjunct Professor at the School of International Service, American University in Washington, D.C.

Readers interested in obtaining FCH case material not included in this book can request copies by writing to the author at 8323 Still Spring Court, Bethesda, MD, 20817, U.S.A.